Safe Computing is Like Safe Sex

You have to practice it to avoid infection

Also by Richard G Lowe Jr

Non-Fiction

Real World Survival: Preparing for and Surviving Disasters
Prejudice and other Irrationalities: Human Rights and Happiness
How to Self Publish: Getting your Book out There (2015)
Sins of the Internet: Electronic Plagues (2015)
Etiquette in the Electronics Age: Turn off the Dam Cell Phone (2015)
I Я A Manager: Managing from the Trenches (2015)
Safe Social Networking (2015)
Cloak and Dagger: Controlling your Online Persona (2015)
Conquered Fears (2016)
How to Promote Your Book: Gaining Eyeballs (2016)
Dying: A Love Story (2016)
Belly Dance Photographer (2016)

Fiction

Fur Baby: Adventures with Humans (2015)
Ghost Healer (2016)
Unanticipated Consequences: Paradox in Time (2016)
Conspiracy: Unleash Hell (2016)
Hells Bells: Hiking Through Hell (2016)
Eight Jewels (2018)

Peacekeeper Series
Peacekeeper (2016)
Battle of Bernard's Star (2016)
Treason of the Admirals (2017)
Neutron Star (2017)
Black Hole War (2018)
Invasion Earth (2018)
Earth's Revenge (2019)
FTL (2019)
First Contact (2020)
The End of Empire (2020)

Richard G Lowe, Jr

Safe Computing is Like Safe Sex
You have to practice it to avoid infection

Published by The Writing King
www.thewritingking.com

Publisher's Cataloging-in-Publication data
Lowe Jr, Richard, 1960-
 Safe computing is like safe sex : you have to practice it to avoid infection
/written by Richard G Lowe Jr
p.cm.
 Includes index.
ISBN: 978-1-943517-00-8 (Paperback)
ISBN: 978-1-943517-01-5 (Kindle eBook)
1. Computer Security. 2. Home computers. 3. Windows series/ I. Title.

This book is dedicated to L Ron Hubbard, the best friend of mankind.

Important notes

Windows focus

This book is focused on the Microsoft® Windows operating system. Many of the concepts described within, especially best practices, pertain to all operating systems. I plan to publish future books in the series about MAC OS X, Android and IOS.

No affiliate links

There are many links to products and services within this book. Because I feel it would be a conflict of interest to earn money for recommending a product, there are no affiliate links within this volume. I will not receive any form of compensation if and when you purchase products that I recommended.

More information

If you'd like more information about computer security, visit the Home Computer Security blog. I regularly post short articles about home computer security issues. You are invited to comment upon those posts; I welcome your input.

http://www.leave-me-alone.com/

Be sure to subscribe to the newsletter by entering your email in the subscription box at the top of the website. By doing so, you'll receive timely information about security and tips on how to keep your computer safe.

Write the author

Please feel free to send an email with any questions or comments. I will try to respond to email messages as quickly as possible.

rich@leave-me-alone.com

Write a review

If you found this book to be of value, please do take a few minutes to leave a review on Amazon telling future readers how you benefited. A good review is highly appreciated and helpful.

QR Codes and short URLs

Some of the URLs in this book are very long and obscure. I've included a short URL using the Google Short URL service with each one. This shortened URL will automatically be expanded to the longer version. In addition, each URL includes a QR code, which you can scan with your mobile phone or tablet to retrieve the exact URL.

Acknowledgements

Many people throughout my career influenced me to write this book. As with everything in life, without the help, kindness, and understanding of others I would not be in the position I am today.

I would like to thank Mr. Henry James, my college professor who started me on the path towards a career in the computer industry. He helped me uncover my hidden talent and love for computing. He granted me the key to the world of design, analysis, and programming, and because of him I began my long climb to a successful and lucrative career.

Also, I owe Steve Davis a debt of gratitude for his patience and understanding in his role as my supervisor at Software Techniques. Steve took me under his wing and guided me through those difficult early years, inspiring me to continue despite the difficulties and challenges. He was my first mentor, and many of the techniques I learned from him served me well throughout my later career. Steve showed me the value of ethics and integrity because those are part of every fiber of his being.

I also thank my good friend Eli Gonzalez, the president of The Ghost Publishing, for his belief in me as I moved into a new career of professional writing. His knowledge of the art and science of the power of the written word is inspirational to me, and motivated me to put on the cloak and hat of a true professional writer. http://theghostpublishing.com

Jimmy James, my peer during my years at Trader Joe's, has been a close friend for the twenty years I worked at that company. We worked together through the good and the bad times, building the computer infrastructure of Trader Joe's from nothing into the powerhouse which runs a multi-billion dollar company. Jimmy has been an inspiration to me throughout my tenure and I admire his sense of ethics and his innate understanding of computer security.

My sister Belinda and her husband Ken Schmahl deserve my special thanks. They operate and own a charter school, The Schmahl Science Center in San Jose California, and I learned much from their example of the value of hard work and dedication. They helped many students over the decades and stand not only as a shining example to me, but to their entire community. http://www.newprod.schmahlscience.org/

Ken Cureton has been my friend and colleague for years, through the good times and the bad. He has remained one of my closest friends for all these years. This man has strongly influenced my life with his sense of humor, honor and integrity, and his incredible intelligence that he has put to good use.

I would also like to thank Steve Levinson, the Managing Director at Online Business Systems. For many years Steve led Jimmy and I through the potential minefield of PCI compliance (the set of rules that must be followed to ensure credit card data is secure.) With his understanding of not just the rules, but the meaning behind the rules, we passed the PCI audit every single year.

Retrosleep, a freelance artist, created the cartoon graphics for this book. You may find him at https://www.fiverr.com/retrosleep.

The cover art for this book was designed and created by Crownzgraphics at crownzgraphics@gmail.com.

Table of Contents

Forward

Steve Levinson

https://www.linkedin.com/pub/steve-levinson/0/904/217

For over a decade, Richard and I worked together to ensure that Trader Joe's adequately protected credit card data from hackers and criminals. It was a pleasure to work with Richard as he was always extremely conscious of the need for security. He was usually on the same page as me and Jimmy James, the Director of Networking.

Each year my team and I performed Trader Joe's PCI (payment card industry) assessment, which was basically an audit of the security controls pertaining to protecting their credit card data. As Managing Director of OBS Global's security consulting practice, and formerly PCI Practice Director for AT&T Consulting, I've performed hundreds of security assessments/audits. In addition to assessing 'current state of being', I've worked closely with our clients to help ensure that they've created a robust sustainable security program. Richard always had a clear understanding that security is an entity into itself and that it requires constant due diligence and proper care and feeding. Our consulting philosophy is to get to know our customers and their businesses very well so we are in a position to not only measure their security, but to also provide meaningful advice.

During my ten years of having worked with Richard, he had mentioned that he had a dream of becoming a writer, and one of the first books on his list was about home computer security. During those long hours of working together on the audits, we had many conversations about security philosophy and various approaches.

I was happy when Richard informed me he had written this book, and honored when he asked me to review and write a forward. It seemed to align quite nicely with his knowledge, wisdom, and experience. After reviewing the book, I can see that Richard has put in a lot of hard work and effort into the work. His list of best practices are spot on, and those of you who embrace these practices will undoubtedly improve your security posture.

This book will prove to be useful and informative to anyone who uses a computer, especially if you don't know much about computer security. It is written in a manner that is understandable and digestible, which is a welcome treat for those of you who have ever been subjected to trying to read through technical books.

Jimmy James

https://www.linkedin.com/pub/jimmy-james/10/721/a97

While I served as the Director of Networking and Information Security at Trader Joe's, Richard was always talking about writing a book to help the home computer users keep their computer secure. I could see that he was passionate about the subject and believed that he could help home users secure their system, but the book never seemed to get off the ground. The fact of the matter was that we worked for a very large retail chain that had stores all over the country, and this required a lot of our time and energy. Being responsible for managing critical and complex areas of the IT department we were required to work all hours of the day and night, including weekends and holidays.

Richard had a demanding yet very fulfilling position at Trader Joe's, and I could see that he loved being able to contribute to the operation and growth of the company. I could also see that Richard truly believed he could help people with their home computer security and I knew he was sincere about writing a book on the subject. I lost hope that he would be able to finish his book project anytime soon.

When Richard finally retired in October, 2013, he said he was finally going to have time to write the books that he'd been talking about for years. In fact he talked about his dreams and plans for writing and publishing over a dozen books on computers, disaster planning and other interesting subjects. He also had a desire to write a few novels and short stories for publication.

Thus it didn't surprise me at all when Richard called me up and asked me to write a forward for his newest book, *Safe Computing is Like Safe Sex: You have to practice it or you could be infected.*

I enjoyed reading this book. It has a different approach than most other books on home computer security. Other manuscripts I've read focused on technical hacks to the operating system, purchasing equipment, and installing products.

Richard's approach comes from our years of enforcing and practicing sound security at Trader Joe's. We both learned that procedures are far more important than fancy hardware or computer programs.

Thus, I was thrilled to find his focus is to teach home computer users to literally practice safe computing. He lays out a series of best practices which, if followed, will dramatically improve computer security.

I enjoyed reading this book, and to tell you the truth, I will spend some time implementing some of the suggestions on my own home computer.

And finally, Richard, I want you to know that I never doubted that you would complete and publish this book. Now I expect you to get going on all of the other ones you and I talked about and get them done as well. I think you have found your true love.

Ken Cureton

https://www.linkedin.com/in/kencureton

Oh boy. Write a forward for Richard's new book? I'm excited to help! Most folks know him as a photographer, which is just one of his hobbies. I can attest that he's a great writer as well!

Richard and I go way back. I mean way back, to the early 1980s. Further than I like to think about. He was working at a small computer company called Software Techniques and my company, Command Computer Systems Incorporated, had contracted with him to write some custom software to allow our computer programs to access data over a network of computers—something that we take for granted nowadays, but very cutting-edge back then.

In those days, Richard was always serious, overly serious. On the other hand, he had a biting sense of humor, which is one of the reasons we became friends. His humor is well, humorously sarcastic. Really, he's a funny man when he wants to be.

The book is very, very good. If you are a home computer user you need to know the dangers that you face every day you venture out onto the internet. It's a wild place, full of hackers and thieves and trolls... no wait. No trolls. Oh yes, there are trolls, aren't there?

Okay, so I met Richard while I was the Vice President of a computer company called Command Computer Systems Incorporated. I'd better mention how easy he was to work with as well as his fantastic personality... sorry, you said no sarcasm? Oh, sorry. Where was I?

Oh yes, Richard was the Vice President of Consulting at Software Techniques. We did a project together and it went very well. I was impressed because he was thorough, the project was completed on time, and within budget. Oh, and his software actually worked as specified. How often can you say all of that?

Over the years he and I stayed in contact and our friendship grew. I remember when Richard got married, out of the blue, after a two week courtship. Twelve years later Richard was a widower, having remained with his wife, Claudia, through a ten year chronic illness. Now that's integrity.

Now he's written a book. I know he's passionate about this subject, and he certainly has the experience. He worked for 20 years at Trader Joe's, and was one of the people in charge of their computer security. That's a tough position, and I know he did it very well.

Buy this book. Read this book. I recommend it without reservation. The knowledge you gain will help you keep the information on your computer secure and safe from evildoers. And to all the evildoers out there—beware! Richard is telling everyone all about your secrets. ALL of them!

Preface

I cried with my friend as we tried, in vain, to recover more than 100,000 personal photos that had been stored on her crashed disk drive. These were photos of her childhood, her teenage years, and her family. Some of the lost pictures were the only surviving record of her deceased mother. We worked long into the night, but the disk was unsalvageable and the photos were lost.

If only she had made a backup of her disk drives.

My coworker screamed when his computer demanded a ransom to restore his files. He paid the ransom but the files were never restored and he lost everything on his computer.

If he had kept his system patched the ransomware would not have been able to install on his computer.

I've been asked hundreds of times during my computer career to help friends, family, and coworkers recover their systems after some kind of disaster. I've successfully restored files from crashed disk drives, cleaned more than 400 viruses off a workstation, and recovered computers after they were taken over by ransomware.

Recovering a computer system after it has been infected by a virus or the disk drive has crashed is hard work. Worse yet, it's often not very successful. It's always gut wrenching to see a friend lose all of their data. It is far worse to find that it cannot be recovered.

This book is a work of love. I enjoy helping people, but recovering a system from these kinds of disaster is not fun for anyone involved.

Thus I decided to write about how to prevent those disasters from occurring in the first place. These pages come from my 35 year career in the computer industry, going back to the original TRS-80 and PDP-11 computer systems in the early 1980s.

I've wanted to write this book for years but given the demands from my job, Director of Technical Services and Computer Operations, I could never find the time to get it done. That job was very demanding, often requiring working late into the evening, on holidays and weekends, and even during vacations.

When I retired after a 20 year career at that wonderful company, I had plans to become a professional writer. That has always been a dream of mine. It's something I wanted to do with all my heart, but kept putting it off until later.

This is the first of many books that are in various stages of being written. As I've moved forward in this writing career, I have found it to be enjoyable and fulfilling.

This book is especially important to me, as by using the information in it, people will be able to improve their computer security. Even if only one person starts making backups and is able to shrug off a hard disk crash, then it was worth all the effort.

I hope this information will be of value.

Introduction

Has your computer ever been infected by a virus? Have you ever felt that sinking feeling when your system wouldn't boot up and you wondered if you just lost all your files? Did you ever find yourself locked out of your computer with a demand for payment or else you'd never see your files again? Even if this hasn't happened to you, I'll bet you know someone who has had these kinds of experiences.

I've helped countless friends and acquaintances throughout my career recover their computer systems from all types of disasters, including everything from a simple virus to attempts to extort ransom. The worst damage I encountered, back in the days of Windows 7, was a computer that had 496 virus infections at the same time. The owner didn't even know his computer was infected! He bought it to me because it was "running a little bit slow."

Computer security has been an interest of mine since my college days in 1981. I fell into the subject by accident. I was attending junior college and I had a hole to fill in my schedule. The only class available was this weird sounding course that involved computers. It looked boring and had nothing to do with my major, but there was no choice. I had to take a computer class and I wasn't thrilled about it at all.

It surprised me to no end when, after just a few days, I discovered that I had found my calling. I loved writing applications (we called them programs back then) to do math problems, play games, display graphics, and keep track of things. I soon realized that I had a certain knack for looking at a problem and figuring out how to get the computer to solve it for me.

The computers we used back then bore no resemblance to the systems of today. The one I used in college was the size of four refrigerator boxes and required heavy-duty air conditioning and power, as well as constant care and feeding from technicians. I was fascinated watching them go about their tasks and talking about this mysterious thing called an operating system.

I decided I had to get a look at this operating system, whatever it was, because it sounded so interesting. One day I sat down at a computer terminal and started trying different password combinations for the administrator account, just for kicks. I had this idea that if I could get into the computer as an administrator I could learn more about how it works. In other words, I was curious.

That was my first and only attempt to hack into a system. It didn't go well because my teacher discovered within minutes what I had done. I thought I was going to get suspended or something worse, but he had a far more devious plan in mind for me.

"Since you hacked our computer you are now in charge of system security," Mr. James, the man in charge of the computer department, told me. "It is your job to keep people out."

Ironically, that's how I got my start in computer security.

As time went by I was hired by a startup company called Software Techniques, and before long I was promoted to the Vice President of Consulting. I worked at that company for six years, then was hired by Beck Computer Systems again as the Vice President of Consulting. In 1994 I moved to Trader Joe's, a nationwide chain of grocery stores, with the title of Director of Technical Services and Computer Operations.

At all of those companies I was the person in charge of the computer security for the company business. That works out to a total of almost 35 years working to keep computers safe from evildoers. At Trader Joe's the responsibility was so large that the job was split between me and the Director of Networking, Jimmy James. I was responsible for the security of the corporate computer systems while he was responsible for all network security.

During my 20 years at Trader Joe's, I worked on hundreds of laptops, desktops and workstations, plus tablets and smart phones. Security was a huge concern as we handled millions of credit card transactions on a regular basis. It was vital to keep those transactions and card numbers from falling into the wrong hands.

I've seen everything that can happen to a computer, from massive virus attacks to worms to physical destruction from fire or water. I've cried along with my friends as they lost thousands of photos forever after being infected by a particularly nasty virus, and helped another friend recover their hard drive after they dropped it out of a second story window.

During those years I helped many people with security problems on their home computer systems. One fact has become clear during all this time. Most individuals believe their computer is safe and secure straight out-of-the-box and are often surprised to learn that, regardless of the operating system or manufacturer, the truth of the matter is quite the opposite.

My goal with this book is to help you fill in those gaps in your Windows® computer security. This will let you focus on using your computer instead of worrying and, worse yet, recovering or rebuilding from disaster.

I hope by passing along my security knowledge you can avoid the types of disasters I've seen throughout my career.

Practice safe computing

In the early part of 2015 a major health insurance company, Anthem Blue Cross, reported that hackers stole the personal records of over 80 *million* customers. Earlier in the year eBay asked more than 145 *million* customers to change their passwords because their accounts may have been comprised. Target reported that the personal data of 40 *million* customers was stolen. Hundreds, if not thousands, of companies of all sizes have been hacked.

"The only truly secure system is one that is powered off, cast in a block of concrete and sealed in a lead-lined room with armed guards - and even then I have my doubts."

– Gene Spafford

If huge corporations with multi-million dollar computer security budgets can get hacked, what hope is there for the ordinary person? If you are anything like me, you don't have huge amounts of money and time to spare to protect your computer. What is the average person to do to keep their systems safe?

I wrote this book to help the average Windows computer user who wants to protect their computer. You can practice safe computing without using your life savings. You don't need to get a university education to understand the concepts, and there is no need to lose your weekends twiddling with your computer system.

Everything presented in this book meets the following criteria:

- Simple to install (exceptions are marked "Advanced").
- Care and feeding is (or can be) automatic. There are a few things which cannot be automated, and these are clearly marked.
- Inexpensive or free.
- Industry standard and supported.

In addition, everything I recommend is something I use on my own personal computers day in and day out. These are solutions that have endured the test of time, and that have proven to be workable and straightforward.

This book is not a tutorial on how to use your computer; rather, it contains tips and techniques which, if followed, will make your system more resistant to hackers and data thieves.

The assumption throughout these pages is that you can navigate around and know in general how to operate basic computer functions. Some tasks you will need to know as you implement these tips include:

- How to navigate to and use the control panel applications.
- How to browse the web and use email.
- How to install and uninstall an application from your computer.

Some of the recommendations are optional, depending upon your needs and your desire for safety. If you only use your computer to browse the web, and don't store any valuable documents or photos, you may only need basic security. However, if you are like most of us, your computer is home to many things:

- Personal photos going back years or even decades.
- Tax documents.
- Letters you've written.
- Faxes you've received and sent.
- Spreadsheets, Word documents, PowerPoint presentations.
- Videos.
- Graphics.

In fact, the data which resides on the computer is often far more valuable than the computer itself. You can always buy another computer, but how difficult would it be to replace your digital photos or the documents you've been writing for years?

Every year Ron put off his taxes until the last possible moment, and this April was no exception. Ron worked late on his taxes, looking through all of the scanned images of each of his business receipts. His computer suddenly crashed and would not reboot. Ron had to hire a specialist to remove the viruses and other malicious applications from his system. Unfortunately, all of the scanned receipts had been destroyed by the virus.

In the interests of keeping it simple, alternatives and more complex solutions are not discussed in any detail. The intention is not to dive deep into the complexities of computer security nor to provide an education in cyber warfare.

The goal is straightforward. If you follow the instructions laid out in this book, your computer will be reasonably secure. You can use the web, email, and your other applications while connected to the Internet, yet still have confidence that nefarious hackers probably won't destroy your system or your data.

I should point out, however, that the only totally safe system is one that is turned off and buried in a hole in the ground, completely disconnected from the Internet and any other networks. Of course that doesn't work in real life.

There are risks associated with connecting to the Internet, and you cannot completely eliminate those dangers. You can, however, follow a few simple rules, install some inexpensive or free products, and reduce your exposure dramatically.

You will learn techniques that you can implement in a single evening or two to secure your computer and your data from all but the most determined enemy. Some of the things you will learn about include the following:

- Building a layered defense.
- Who hackers are and what they want.
- Choosing good passwords.
- How to keep your data safe from any kind of disaster.
- Keeping your system up-to-date with the newest security corrections.
- Safely browsing the web.
- Getting rid of advertisements.
- Antivirus solutions.
- Building a wall (firewall) around your system.
- Securing your network.
- Securing your wireless network.
- Protecting your email.
- Protecting your children.
- Protecting the devices on your computer.

The emphasis of this book is on solutions rather than theory or technical information. For those who want a deeper understanding of security, I've listed some good books in the references.

How the Internet works

This book is intended to be understood by anyone who has some experience using a home computer. I've tried to avoid using complex computer terms except where absolutely necessary, but there are some concepts that require specialized words.

You should review this chapter and ensure you understand all of these words and concepts before going on. These will be used throughout the book because this is the basic terminology used to describe the world of computers and the Internet.

A **device** is simply a piece of equipment which, for the purposes of this book, is connected to your computer or your network. Your desktop computer is a device as is your digital camera and your printer. Your smartphone (regardless of the model), tablet, video game consoles, scanners, and even smart light bulbs are also devices. Any computer, anything that can connect to a computer or anything attached (or that can attached) to your network is a device.

A **resource** is any device or thing that can be used by a computer to serve some function. A disk and a flash card are resources. In addition, a folder which you have shared (made available to others) is a resource. Printers, scanners, game consoles, and similar devices are also resources.

Networks

A **network** at its most basic is two or more computers that are connected in order to share things (devices and resources). A simplistic definition of the **Internet** would be that it consists of all of the networks in the world. This is not strictly true as many networks are not connected to the Internet, but the definition is useful for this discussion. Many use the term **web** (also known as the **World Wide Web**) interchangeably with the word Internet, but they are not the same thing. The web is the part of the Internet which is accessed using web browsers such as Internet Explorer, Google Chrome, Firefox, Opera, or Apple Safari. The Internet is the web plus a whole heck of a lot more, including email, **FTP** (a method for moving files around the Internet) and other similar things.

You might have a network in your residence which consists of a desktop computer in your home office, a wireless laptop for the kids, a tablet for your books and an Xbox® to play games. Quite often networks will also include printers, hard drives, scanners, cameras, and other devices.

Why would you want to have a network in your home? In order to **share resources**. Some typical resources include:

- Printers.
- Scanners.
- Fax machines.
- Computers.
- Laptops.
- Tablets.
- Smart phones.
- Folders on a computer.
- Disk drives.
- Game consoles.
- Cameras.
- Smart light bulbs and other intelligent devices.

By grouping these devices together in a network, anyone who can access the network can use those devices (this can be restricted by various forms of security on each device). Thus you can print from the computer in your home office to the printer in the basement, or download a game onto your laptop and move it to your game console.

Computers

A **server** is a computer that serves a purpose or function. When you use your Xbox to play games over the Internet, the game console is communicating to a server (actually many servers in what is called a **server farm**) which provides gaming services. When you use online banking to balance your checkbook you are using the banks server to get this work done. There are many millions of servers doing a huge variety of tasks all over the world.

On your computer you have available to you a number of **applications**, each of which allows you to perform one or more functions. You might use the Quicken application to balance your checkbook, ITunes to play music, and Google Chrome to surf the web (web surfing is also known as **browsing**). Each of these is an **application**. They are also called **programs**, and the two terms may be used interchangeably.

A special kind of application intended to do something technical for your computer is called a **utility**. Microsoft Windows has a number of utility programs called **control panels**. Your computer comes with many other utility programs, some of which are mentioned in this book and many others that are not. You should make sure that any utilities you download off the Internet are from reputable sources and have been checked for viruses using the techniques described in the section titled *Safely downloading files from the Internet*.

The **operating system** is the application that administrates (operates) and tells the computer what to do. Windows is the Microsoft operating system and it works on PCs, tablets, and other devices. Android is the operating system created by Google, IOS is for Apple iPhones and other devices, and the MAC OS X is for Apple computers. There are many other operating systems, but the average user is unlikely to be using them on their home computer.

There is yet another type of application called **firmware**. These are programs which run directly inside a device to make that device function. Thus cameras, printers, disk drives, and every other device has its own special firmware which actually operates the device itself. Your personal computer has its own firmware to control the computer hardware.

MAC address

Every single device that can connect to a network is assigned what is called a **MAC address (Media Access Control)**. These addresses, which are simply long numbers, are assigned by hardware vendors when the device is manufactured. They are unique and never change. Sometimes computers have more than one MAC address. This happens when they have more than one network connection. Each connection has its own MAC address. For example, if your computer has a wireless connection and can also be connected with a wire, then it will have two MAC addresses.

This is an example of a MAC address:

00-FB-60-D8-18-2B

The purpose of a MAC address is very simple. Imagine a town with a few hundred houses. The land of the town is divided up into lots of land. Each lot has a plot number which identifies the exact lot. These unique numbers allows each lot to be identified on tax and property documents. The plot number, like the MAC address, doesn't change. The MAC address is similar to a plot number in that it uniquely identifies something. The first six digits identify the vendor and the second set of six digits identifies the device.

IP address

Each device that can connect to a network is assigned an **IP address**. The lots that we mentioned before include homes, shops, factories, and other places. Each of these is assigned a street address. An IP address is, in concept, the same as a street address. It is used to identify a logical location. The post office, for example, uses your street address to know where to deliver your mail. Likewise, an IP address identifies where to deliver your print job.

A typical IP address is shown below.

192.168.10.15

The IP address shown above is an **internal IP address**, which is also called a **NAT address**. IP addresses were created decades ago, long before the Internet or the web existed. The IP address format allows for a little over 4 billion devices to be connect to a network. That might seem like a lot, but think about how many devices you have in your home. You might have a desktop computer, a laptop for the kids, a tablet, an Xbox, a PlayStation, an iPhone for each family member, and so on. Each and every one of these requires an IP address. With more than 7 billion people in the world, there are not nearly enough IP addresses for every possible device.

There are some additional complexities that make the total number of useable IP addresses even less. The point is there are not enough IP addresses by any stretch of the imagination to handle all of the devices throughout the whole planet.

Note these examples discuss IP4, which is the original IP address format. In order to solve the problem of running out of IP addresses, a new format called IP6 has been created. The internet is slowly changing over to use the IP6 format.

Let's back up a minute and discuss how your home network connects to the Internet. Your Internet provider supplies you with a device which they call a **router** (they might tell you it is a cable or DSL modem or something similar.) You can think of a router as a switchboard of sorts. Let's say you have a FIOs modem (which we will refer to as your router) which was installed by your local cable company. When you connect to a web site on the Internet from your laptop, your computer is actually talking to your router. The router then figures out the best route (hence the name router) to get to the web site you asked for. Sometimes your home router communicates with the routers at your internet providers, who then figure out how to get to wherever you are trying to go.

The router has its own IP address, which is *not* an internal IP address. A router's IP address might look something like this:

$$3.18.231.42$$

You could say the router has two sides. One side, which is the cable connected to the wall or telephone jack, faces the Internet. The other side connects to your internal network.

The Internet-facing side uses an IP address like the one above; this address is also known as a "public" IP address, since it is publicly-facing. This address allows it to communicate with other systems and routers on the Internet, and them to communicate with it.

The side which faces your internal network uses an internal IP address. Thus each home network uses only one IP address which is visible to the Internet. All of the internal IP addresses are *not* directly visible on the Internet.

Each time your computer is attached to your network it asks your router for an internal IP address. This is done using **DHCP**, which is the technical standard for doing this. When someone refers to a DHCP address, it means the same thing as an internal IP address. This implies that the internal IP address of your home computer, and every other device attached to your network, can change from time to time.

So to boil it all down, your router has a permanent address (or semi-permanent since a router's public IP address can change occasionally) which identifies it on the Internet. Devices within your network, on *your* side of the router, use internal IP addresses.

This is extraordinarily important for the security of your computer system and all of the devices within your home or office network. Since all communications are done by first communicating with your router (which, remember is your cable, DSL or FIOs modem), it means that no device on your network is directly communicating with anything on the Internet.

The reverse is also true. No device, no machine, no hacker, no person, nothing at all can communicate directly with anything on your home or office network (note there are exceptions to this rule, but for the purposes of this simplified discussion this is true.)

What this implies is your router, simply by assigning internal IP addresses to your computers, acts as a kind of barrier between your network and the outside world. Your router is literally your first line of defense in your security system, even if it does not include any other security at all.

We're almost done with definitions, but there is another very important piece of this puzzle to understand. The web and the Internet don't communicate using names, they use IP addresses. Luckily we humans don't have to memorize IP addresses when trying to find a website such as google.com. This means that something must perform a behind-the-scenes functions to translate it into an IP address.

DNS

Translating names into IP addresses is done using a function called **DNS**, which stands for **Domain Name Service**. A **domain name** is a name such as google.com, microsoft.com, or leave-me-alone.com. When you type a name into your web browser, it asks DNS to translate that name into an IP address. You could think of it as a kind of phone book.

In this case, the phone book is located on another computer somewhere out on the Internet. Again this is a simplification of what can actually be a complicated process.

DNS opens that phone book, looks up the name which you typed, and finds out the IP address. To make everything go a little faster after that, DNS typically makes a record of that IP address and name on your local system so it doesn't have to go back out to the phone book again.

In a nutshell, when you enter a name into your web browser (Google Chrome for example) that name must first be translated to an IP address. Your browser does that for you using DNS. Once the browser knows the IP address, it reaches out over the Internet, through your router, to get the desired web page.

This is a bare bones, heavily simplified description of how your home network and your computer operate on the Internet and the web.

Defense-In-Depth

Merely connecting a computer system to the Internet exposes it to all types of attacks. Your computer can be attacked or raided from individuals in countries 10,000 miles away or from next door. You can infect your system just by opening an email, clicking a link, or viewing a web page. Even plugging a brand new camera into your system to view your pictures can result in a viral attack.

Considering that attacks can happen at any time and can come from anywhere, you would think that every computer has some kind of built in, bulletproof security. While it is true that all operating systems have some kind of protection against attack, often it is woefully insufficient. Even in the cases where security is well engineered it becomes obsolete in a very short time because the Internet and the nature of the threats changes constantly.

There are many different ways a computer may be attacked. It is important to understand that all of these attacks can occur on every model of computer and operating system that exists. No manufacturer, no operating system, and no computer is immune. Some of the more common attack methods include:

- Clicking on a link which installs a virus.
- Reading an email which installs a virus.
- Viewing a web page which installs a virus.
- Getting infected by a virus buried in an Internet advertisement.
- Hackers directly attacking a system through your network.
- Hackers who create a wireless hotspot which they control to steal your information.
- Deceiving you so you click a link or install a program to hijack your system.
- Plugging in a camera or flash drive which someone has infected with a virus which then gets installed on your computer.

Although many security products claim they protect against all of these various threats, none of them is entirely effective. This doesn't mean you shouldn't use one of the major Swiss-army-knife type security applications; it means you should supplement their protection with other products to fill in the gaps and make your system more secure. Keep in mind that whatever products you use to protect your computer are also purchased by attackers who test them in their labs looking for way to circumvent these controls.

Think of how difficult it was hundreds of years ago to defend a castle. The fortress had to be protected with a high wall to keep invaders out, and the wall needed portals for defenders to shoot arrows and other weapons from without exposing themselves to return fire.

The castle needed to be protected against battering rams and rocks launched from catapults. The doors and other wooden parts required some protection from flaming oil, and the defenders even had to consider that soldiers might attempt to dig *under* the walls to cause them to collapse.

They also had to protect against Trojan horses, which are attacks disguised as something good or useful, traitors, and even unknown types of weapons. They had to have a contingency plan ready for the possibility of the attackers penetrating the walls of the castle.

How did these medieval engineers protect these fortifications from all of these types of attacks? I know this appears to an impossible task, but believe it or not, engineers designed and built castles in Europe that survived attack after attack for hundreds of years. It wasn't until the invention of gunpowder and canons that castles finally met their match.

Military engineers created what's called a *defense in depth* or a *layered defense*. A typical castle consisted of a fortress, called a keep, surrounded by a thick inner wall, in turn surrounded by a high outer wall which may also have been surrounded by moat full of water. They covered the doors in steel so they couldn't be burned and built a walled courtyard within the castle to protect against invaders who breached all of those defenses. They had buckets of boiling oil to pour on troops attempting to climb up the walls and narrow slots for soldiers to shoot arrows from.

Security for your computer system works the same way. For example,

- A firewall is literally a wall, albeit electronic, between your network and the Internet that help keep the bad guys out.
- You could think of antivirus software as soldiers which search through your castle for any enemies that made it through your defenses.
- Passwords are, of course, required to get past the guards and into your castle.

The idea is to install applications on your computer and to configure them properly so you can rest easy, knowing no matter what happens you are protected. In fact, you even need to be prepared for the absolute worst possible case of a system crash or a nasty virus which destroys your data.

It is important to understand that installing security applications is only half the picture. You also need to be following what are called **Best Practices**. These are generally accepted recommendations of how to best secure your system from various types of attacks.

> **Best Practice**
> *Follow all best practices for security to ensure your computer is as safe as possible.*

For example, in medieval times everyone needed a password to get into the castle. This was their best practice: always ask for the password and get the correct answer. They would also make sure they changed the password occasionally and made it difficult to guess. As you use your computer, you need to follow best practices to ensure your safety. These should become part of your everyday routine.

There are several products that can help to make this simpler, and I will be presenting them to you throughout this book. Just remember that violating your best practices can make your system insecure and let the bad guys in, even though you have the absolute best security applications installed.

All best practices will be clearly noted and explained.

In the next sections we will look at each part of your computer, including your operating system, hardware, network, wireless connections and everything else. We'll discuss applications that can help tighten your security, and best practices you should follow to help keep your system safe.

The enemy

Hackers

I'm sure you've heard the word **hacker** a few times, perhaps in the news or in a conversation with that weird tech guy at the computer store. You possibly have a picture in your mind of some awkward pimply teenager who sits locked in a room and plays on computers because he's afraid of girls.

> **"Hackers are breaking the systems for profit. Before, it was about intellectual curiosity and pursuit of knowledge and thrill, and now hacking is big business."**
>
> **— Kevin Mitnick**

While I am certain those people exist somewhere in the world, according to the Merriam-Webster online dictionary, a hacker is defined as "a person who secretly gains access to a computer system in order to get information, cause damage, etc.: a person who hacks into a computer system."

In the old days before the personal computer was invented, hackers were often high school or college students who were curious about computers and had few social skills. They broke into systems because it was fun and a learning experience (or so they usually claimed.) Occasionally these people committed minor crimes such as stealing information or crashing computer systems, but in general they were not so much criminals as lonely, socially awkward people.

The world has changed much in the fifty years since those days. Today hacking is a huge criminal business, often associated with organized crime in foreign countries such as the Czech Republic, Romania, Bulgaria, and Russia, to name a few. In years past, hackers were mostly a nuisance; today it's possible for them to crash the stock market, destroy industry, and interrupt telephone and power service.

In other words, hacking has become big money. You might wonder with huge targets such as banks, military installations and the like, why a hacker would be interested in your little home computer system. There are two major thrusts to their needs.

First, they want to directly attack your system to steal something from you. Some of the things they do once they have hacked into your system are listed below.

- Steal your usernames and passwords.
- Steal your identity so they can hijack your credit.
- Steal credit and debit card numbers.
- Add themselves to your accounts so they have control.
- Sell your information and credit cards.

Some hackers, though, want to take control of your system to create what is called a **bot** or a **zombie**. They want your system to become their slave or robot (hence the word *bot*.) If they are successful, your system can be used for any purpose they want, including:

- Attacking other computers.
- Breaking into other systems.
- Attacking the networks of companies and other nations.
- Storing files, including pornographic images.
- Or just about anything else they desire.

These hackers, if they successfully break into your system, will add it to a collection, or network, of hundreds, thousands or in some cases even millions of bots. These are called **botnets**. Criminal hackers rent out parts of these botnets to perform tasks for other criminals.

Let's say someone wanted to send out 100,000 spam emails without being detected. They could rent a thousand bots from a botnet for an hour for anywhere from ten to a hundred bucks. Those bot systems would then be used to send their spam. This makes it almost impossible to trace the spam back to the originator.

The best recommendation for you is to prevent hackers from gaining access to your system in the first place. If your defenses are penetrated, you need to ensure that you can eradicate the infection or, if needed, erase and rebuild your computer from the ground up.

Some hacker terms

Throughout this book you'll find many words specific to hacking and computer security. Some of those words are defined here since they are used throughout the text. Others will be defined as they are mentioned.

Cracker – A person who breaks into computers.

Cyberwar – When nations use computers to attack the computer resources of each other. The Stuxnet virus, for example, was purportedly created by the United States and Israel to attack the computers controlling the motors of the centrifuges used by the Iranians to create nuclear fuel.

Denial of service attack (DoS) – An attack whose purpose is to make a web site or network unresponsive. A **distributed denial of service attack (DDoS)** is an attack from a large number of computers, typically a botnet.

Hacker – An individual who uses computers in an unauthorized way to gain access to data. Any form of hacking, unless it has been authorized by the computer owner, is illegal and criminal. **Black hat** hackers engage in criminal activities while **white hat** hackers use their hacking skills to help improve security.

Key logger – An application which records every single key that is typed and every mouse click that occurs on a computer. These are often planted by viruses or malware in order to get passwords and other information.

Logic Bomb - This is a virus or piece of code, often installed by hostile employees or contractors, which is timed to trigger its payload at a specific date in the future. For example, a malicious employee might leave some code embedded within the accounting package which causes it to delete everything a year in the future. These types of infections are very difficult to detect and even harder to eradicate, as they may have been added to the system years before. Even backups may be corrupted with the malicious code, so restoring may not be an option for recovery.

Malware – This stands for malicious software. It means anything which intends to do you or your computer system harm, or to use your computer to harm other computer systems or people.

Payload – This is the part of a virus which performs the intended action. This could include adding the computer to a botnet, recording keystrokes, or destroying data.

Ransomware - Some of the more malicious viruses actually hold computer systems for ransom. These often first appear as a popup telling you your system is infected and asking if you'd like install the "antivirus" software. If you click anything what actually gets installed is a very-hard-to-delete virus. This virus demands that you pay a certain amount (usually a few hundred dollars) for the safe return of your files. Some of the more sophisticated varieties actually encrypt your files so they cannot even be used until you pay the ransom. Often the only way to recover from these types of viruses is to restore your files from a backup.

Spyware – This is any application or technology which gathers information via computer without consent. Although spyware is unethical it is often legal because users themselves authorize it to be installed on their system. Keep in mind that the terms and conditions of many applications frequently authorize spyware to be installed.

Time Bomb – This is a virus which triggers at some date in the future.

Trojan horse - This type of malicious code is a virus or other dangerous program which is embedded within a desirable application. For example, someone might post a very nice screen saver on their web site which includes a Trojan horse that deletes files, sends information to a criminal origination, or perhaps just waits for instructions (this is called a **bot** or **zombie**).

Viruses - A virus is defined as an application which can create copies of itself and install those copies on computer systems. Generally a virus tries to remain hidden so it can do what its creator intended without interruption from the user of the computer. Usually a virus consists of two parts. First is the delivery system, an email or web page or something similar, which gets the virus onto a computer system. Once the computer has been penetrated, the virus will install a **payload**, which is the actual virus application.

> **"I think computer viruses should count as life. I think it says something about human nature that the only form of life we have created so far is purely destructive. We've created life in our own image."**
>
> **— Stephen Hawking**

Warez – Versions of applications such as Microsoft Office or Photoshop which have been hacked so they no longer need a license to be used. This is also called **Pirating**, and it is illegal. Warez applications are often made available for download on hacker web sites. You should never visit these web sites, much less download any Warez programs. If you do you will almost certainly find your computer infected with malware.

Worm - A worm is a self-replicating virus. Some of the more common worms, such as "I Love You," use Outlook to send themselves to every email address listed in the contact list. Others actually have their own email system built into themselves so they can send to every email address they can find in any file on the hard drive. Some worms, such as Nimda, actually install themselves on web servers and then search through the Internet for other vulnerable machines. When these machines are found, the worm penetrates them and installs itself automatically.

As you can see, these definitions can overlap. It's possible to write a Trojan horse which launches a worm which then installs a logic bomb which is set to launch itself again a month later and start the whole process over. When you open the infected email nothing happens at all for a month, until the logic bomb triggers. Or a malicious programmer could leave a logic bomb behind him after he leaves a company which launches a month later. This program could scan the corporate address book and send copies of itself to every person listed within. The payload of this worm could be a cool graphic which is actually an application which, when opened, deletes every file on the disk.

Common attacks and what to do about them

Social engineering

> *At the retail chain where I used to work we learned about social engineering the hard way. We found out that hackers would dress up as computer technicians, boldly walk into a store and explain they were dispatched to replace the credit card readers. The idea was to replace those card swipers with skimmers that sent any credit card numbers to the hacker's computer as well as our own. Our stores were briefed to demand a certain type of identification from all technicians so the hackers were unsuccessful.*

The most important tool available to hackers is called **social engineering**. Believe it or not, this is not some fancy application or a super hacker missile. In the days before computers this was called a *con job*. Pure and simple, social engineering is the attempt to con someone into doing something at the behest of the hacker.

If you saw the movie *Hackers* from a few decades ago, you'll remember the scene where one of the hackers dressed up as a computer technician and boldly walked into a business to install special equipment. That was one form of social engineering, where the hacker pretends to be someone he's not in order to gain access to your equipment.

> **"Social engineering bypasses all technologies, including firewalls."**
>
> **– Kevin Mitnick**

Social engineering is defined as tricking people into bypassing or breaking normal security procedures, or, more simply, conning them into doing something having to do with computer systems. You normally wouldn't click a link that said "click here to install nasty virus", would you? But you might click a link that pretended to be a letter from a friend, a notice from your bank, or an advertisement promising vast financial gains.

Attackers use social engineering to get you to relax your defenses so they can install malware themselves on your computer system. A virus might be hidden within a cute screensaver, an application that looks interesting, or a photo or video. Each of these things tries to get you to download them or click a link so the virus can do its dirty deed.

Drive-by attacks

The link promised pictures of cute cats, so Shelby clicked on it right away. She loved animals, especially cats, and spent a bit of each day laughing at funny pictures about them. Shelby had not patched her system in years which made her vulnerable to drive-by attacks. As soon as the web page opened in her browser, Shelby's system was infected with a particularly nasty virus.

Sometimes just opening a web page or email message can infect your system with a virus. All you need to do is visit or **drive-by** that page in order for malicious code to be downloaded onto your computer. These types of viruses can only be successful if you have not kept the operating system and applications up-to-date with recent security updates.

Sometimes these drive-by attacks occur when advertisements are displayed. Criminals might create banner ads which infect your system as soon as they are viewed on a web page.

The following actions will help prevent drive-by attacks:

- Ensure security updates are regularly installed on your system for the operating system and all applications.
- Use a product such as AdBlock Plus to remove advertisements from your web browser.

Phishing

Bill, a friend of mine, woke up late one night and couldn't get back to sleep. He decided to check his inbox since he had nothing better to do. He saw an email that appeared to be from PayPal stating there was unusual activity on his account and he needed to login to recover. He clicked the link, entered his username and his password. Afterwards he went back to sleep, happy that his PayPal was protected from fraud. Unfortunately, he fell for a phishing scheme and actually give hackers his username and password. Within a short time all of the money in his PayPal account was gone.

Spam email messages are often attempts to use social engineering to steal login information. One form of this is called **phishing**. This is when an email pretends to be from your bank, PayPal, the IRS, or something similar and asks (demands) you click a link. The email will look very similar to, if not exactly like an official message from that bank or other company.

Once you click the link a web page that looks exactly like one you would expect will ask you to log in by entering your username and password. Once you enter that information, you will be told the password is incorrect and the web page will be redisplayed. Unbeknownst to you, the first web page was a hacker site. It stole the password and username that you entered, displayed the error message, then passed control over to the correct web page with you none the wiser. The hacker can now log into your account anytime he or she chooses.

Spear phishing

As soon as she arrived in the morning, Sally saw she received an email from her colleague at an old company where she used to work. She was excited to hear from him and quickly clicked the line within her inbox. Since her computer had not been patched in months she was automatically infected as soon as she clicked the link to open the message.

This specialized form of phishing occurs when a specific company or group is targeted. Standard phishing tends to be sent out to a random group of people, while spear phishing is targeted. These attacks can be very specific, with messages discussing relatives, co-workers, competitors, and so on.

For example, let's say you receive an official looking email from your college asking to perform a survey. The questions are included in an attachment which you are asked to download, fill out, and return via fax. If you download the attachment and open it with Word, you could be infected (especially if your system was not completely patched.) This is a spear phishing attack and the attacker did some research to find out the name of your college. The idea is by including specific information the receiver of the message is more inclined to open it and follow the instructions.

How does this work? Someone targeting a company will look into the background of people working at that company. Using that information they will create fake emails, letters, and whatever else is needed to get you, and a number of other specifically targeted people, to click a link, download a document, or run an application.

How do these hackers get information? They scour Facebook® and other social networking pages, LinkedIn® profiles, credit histories, and anything else they can find. They use this information to send you very specific emails pretending to be from someone you trust.

Man in the middle attacks

Nathan was in a hurry. As always he stopped at his local coffee shop in the evening after work to log onto their wireless and get a bit of work done. Today he was happy to find a new hotspot, called "fast coffee shop" has shown up on the list of wireless networks available. He connected to it without much thought and did quite a bit of work. Unfortunately, Nathan connected a few personal sites without using https, and the hacker who had provided the hotspot stole some of Nathan's passwords.

Have you ever connected to the wireless hotspot at a coffee shop or similar location? Have you used a random hotspot which was displayed in the list of wireless networks on your computer? Or have you connected at an airport, bus terminal, or anywhere else?

When you connect to a hotspot such as these, you are vulnerable to what is called a man-in-the-middle attack. Typically this type of attack occurs over wireless connections, although it can happen with any type of network.

Do you remember the old movies that showed someone listening in on a phone conversation? That was a type of man-in-the-middle attack. In that case, while talking on the phone, a third person listens, and possibly records, everything in the conversation. The listener hears everything said with the two other parties on the line unaware they are being spied upon.

There are several ways this can happen to wireless network users. A hacker might install his own wireless router in the building next door to your local coffee shop, airport, bus station, or whatever business is around. He'll give the hotspot an attractive name like *free-wireless* or simply *coffee shop* and let anyone use it.

It's a fact that wireless hotspots are very popular. One set up in the manner described above in a prime location might attract dozens of people to connect in a matter of minutes. The hacker can monitor and record the transactions of each of these people. The hacker's primary intention is to obtain usernames and passwords of bank, email, and other high-value accounts.

In another form of a man-in-the-middle attack, a hacker uses special equipment to listen to the wireless signals in the air. This is called **sniffing** and it can happen anywhere at any time.

How do you protect yourself from man-in-the-middle attacks?

- Always use HTTPS to connect to any web site containing personal or sensitive information such as bank, credit card and email accounts.
- Use a service such as Private Internet Access to secure everything on your wireless connection.
- Ensure you are connecting to a known and trusted hotspot.
- If possible, use a wired connection to the Internet.

Trojan horses

My pager had gone off at 4 a.m. one morning in May 2000. I didn't recognize the number, and when I called back it was the CEO of the company. The email system was super slow, to the point of being unusable, and he wanted me to look at it right away.

After hanging up and muttering a few choice words, I connected to the email server. The machine was going painfully slow. It didn't take long to uncover that our email system was attempting to send hundreds of thousands of messages. We had been hit by a new virus called "I Love You."

What happened is someone received an email with a subject of "I Love You" which contained an attachment. They presumed the attachment was a note from someone they loved so opened it on their computer. The "I Love You" virus was triggered and sent itself, using our email server, to every single person in their address book. Everyone in the corporation received the small email and did exactly the same thing. It didn't take long for the email server to grind to a halt. The time, 4 a.m., corresponded to the time the warehouses and office on the east coast started their work day.

One of the primary missions of a hacker is to convince users to install programs on their computer. To do this, they use social engineering. For example, they might offer a nice screensaver or electronic postcard on a web site. Underneath the covers of these is a harmful or undesirable virus or other malicious application.

This seemingly useful software or product is called a **Trojan horse**. Do you remember the story from the *Iliad* of the war between the Greeks and the Trojans? The Greeks had besieged the city of Troy for ten years and they desperately wanted to end the war and go home. They decided to use a little deception to put an end to this long war, and built a beautiful wooden horse. The inside of the horse was hollow and just large enough for a few Greek soldiers. The Greek army withdrew, pretending to be giving up the battle, leaving the horse behind as an offering to the gods. The Trojans dragged the horse inside the walls of the city. After nightfall, the Greek soldiers crawled out and let in the main Greek army. The city of Troy was completely destroyed.

In the computer world, a Trojan horse works in exactly the same way. It is disguised as something valuable, something that someone would want to install on their system. It might be camouflaged as a screen saver, an interesting document, a useful tool, or even a picture or graphic. The idea is to convince you to let down your guard, download and install something in order to infect your computer.

That was how one of the original destructive viruses, "I Love You," spread so fast. That virus appeared to be something very desirable, a love letter from a friend, so many opened the attachment without evening thinking about it. Once they did, the malicious virus was unleashed to do its terrible damage.

Another tactic is to bury a virus within a useful and attractive product. Since these appear to be valuable, thousands or even millions of people may download them onto their computers. The product generally does exactly what it is supposed to do, but it also installs the virus. The hacker gains control of all of these systems. If the virus is well written those users may never know their systems have been compromised.

Sometimes a Trojan horse can have extremely expensive repercussions. Many web sites, especially those offering pornographic materials, ask you to install an application in order to see their content. This is "free" no questions asked, except once installed, this application scans the computer for a modem or fax line. If it finds one the application can use it to dial a 900 or other pay-per-call number. If undetected, this can result in phone bills of hundreds or even thousands of dollars very quickly.

Many viruses and worms have been very successful using these tactics. On any given day you might receive dozens or even hundreds of emails attempting to get you to open an attachment. The email message is constructed so that it appears to have been sent by a friend. Thus, a naive or new Internet user might open it to see what his "friend" has sent. Even experienced security people occasionally get taken in by this type of social engineering.

Often the Trojan horse is a very small program. It installs itself on a computer, then waits for commands from its hacker master. The hacker can change this application any time he wants to perform new actions.

Back doors

Sometimes an employee or contractor will create themselves a hidden way to get into a computer. Often these are created just out of laziness or expedience, as when a contractor creates an administrator account without permission. Sometimes back doors are set up for malicious reasons, such as when a hostile employee wants to damage a system without leaving a trace. These actions can be very difficult to discover, as they are often well hidden. A back door can be as simple as an unauthorized administrator account or as complex as specialized code added to an operating system. Back doors have been known to be created by vendors to allow their technicians to log in when doing maintenance (which is highly unethical) or by hackers who break into a system and install code so they can log in anytime they wish.

*Many years ago I was the Vice President of Consulting for a small computer company. We specialized in writing applications for small to medium businesses on larger computers. All of the people who worked on my team were good **programmers** (the people who create applications.)*

One of my team members was a man named Dave. During the interview process he claimed he had experience in programming and produced references and samples of applications he had created to prove it. After working with us for three months it was pretty obvious his skills were lacking, so I terminated his employment.

A few weeks later a member of my team noticed our computer seemed to be doing too much work. If you've ever spent many hours in front of one of those old computers, you know that you get used to hearing a certain pattern of activity. The disk drives, magnetic tape reels and other equipment made a lot of noise. The more active the computer was, the louder and more frantic the whirs and clicks became.

I logged into the machine myself to see if anything obvious was wrong. After some investigation, I realized someone was logged as the administrator, which meant they could do anything they wanted on our computer. It took some digging, but after a few hours I discovered that Dave had left behind something called a back door. He had modified the operating system so he could log in as the administrator anytime he wanted.

When had he done this? I found out from my investigation that he had made the changes only a few days after he was hired! I had to personally examine every single program at all of our customer sites for any suspicious modifications; I found one other computer that was compromised. I never found out what he had been trying to accomplish and never knew what happened to him. He just disappeared after I fired him, which, considering what I had discovered, was fine with me.

Safety in life

Have you ever left your computer sitting on the table at your local fast food place while you rush to the counter to get your food? Do you throw bank statements, credit cards receipts, and other pieces of paper directly into your trash? Do you enter passwords and PINs without looking around and without using your hand to keep others from seeing?

If you do any of these things, you are making it easy for criminals to hack your life. You can lose your computer, your identity can be stolen, and your debit and credit accounts could be at risk.

Thus, while you are working on making your computer safe, it is also a good idea to take a look at your life routines. Unsafe habits can very effectively obviate or destroy the security of your computer, your online accounts, your home, your privacy, and even your life.

Lock your devices

Take a look at your smartphone, tablet, or other mobile device. If you are like most people, you've set these up so you can use them to do just about everything needed to manage your daily life. You probably have installed applications similar to the following:

- Email with full access to your inbox and all your other email folders.
- Twitter®.
- Access to all your bank and credit card accounts.
- Online games.
- PayPal®.
- EBay®.
- Storefronts for everywhere you shop.
- Amazon®.
- And many, many more.

It is quite common to set up these applications so they do not require passwords and usernames to be entered each time you access them. It's normal to not enter a password every time you want to read an email message.

Think about what will happen if you lose your phone. I mean besides the loss of the phone itself as well as your contacts, photos, and videos. Think about everything else that is on your phone.

A thief who steals your smartphone could gain access to all of these accounts. By far the most important account is email. If a thief gets access to email he can get passwords to your entire online life. This is done by requesting that those accounts email the password (or a password reset), and since the criminal has access to your inbox, he can gets what he needs to access your information.

Additionally, you may have set up your phone to access your secure wireless network in your home. A thief who has your phone and knows where you live can use this data to break into your wireless network. Once inside your network, a skilled hacker can attack your computers at will.

> **Best Practice**
> *Use a PIN, password or other ID to lock your phone, tablets and other devices when you are not using them.*

How do you keep yourself safe? Lock your phone, tablet, and all of your other devices. Every one of them has a method to require a password, a PIN or, in the case of the iPhone, a fingerprint. If you don't do this, when you lose your device (or it is stolen) you are basically handing over the keys to your life. There are also apps that you can use to cause the phone to wipe itself clean if it is lost or stolen.

Keep a note of what applications you have installed on your phone and other devices. As soon as possible after you realize they are missing change the passwords for every single account that those devices could access.

Do this regardless of whether you locked the devices or not. Locking the device gives you some additional time to change those passwords. But all locks can be defeated.

Shred everything

One of the methods used by hackers and others to break into your computer, your online accounts, and even your life is called **dumpster diving**. Sometimes malicious individuals really do go into trash dumpsters to look for clues to commit crimes. It's interesting how bits of paper with important data can wind up in the wrong hands at the wrong time.

Paper is used for so many things in a person's life. Think about all of the paper you handle constantly as you go about your daily business.

- Cash register receipts.
- Bills.
- Magazine covers with address labels.
- Envelopes with addresses printed on them.
- Tax returns.
- Paycheck Stubs.
- Bank statements and canceled checks.
- ATM receipts.
- Credit card receipts and bills.
- Utility bills
- Insurance information.
- Credit card and bank account statements.
- Personal documents.
- Pictures.
- Warranty information.
- Bits of paper that you've written things down on.
- Military records.
- And much, much more.

All of these documents contain information that may be used to identify you in one way or another. If a malicious individual can get hold of several different documents he or she can steal your identity, break into your accounts, and locate your home or place of business for robbery or home invasion.

Once you throw out a piece of paper you lose possession of it, as well as all legal rights to it. The police may examine anything found in your trash, for example, and anyone can grab anything from your dumpster at any time (as long as the dumpster is not on your property.) Worse yet, that piece of paper will be carted to the local landfill (in most cases) and will remain there virtually forever. While it is probably unlikely that someone will dig through a landfill, find your credit card slip, and use it to hack into your account, stranger things have happened.

> **Best Practice**
> *Shred any and all paper which contains any identifying information.*

There is no need to be paranoid about these documents. You can rationally and easily prevent any printed information from falling into the wrong hands by purchasing an inexpensive cross-cut shredder and using it to chop up every applicable document before you throw it in the trash.

Make sure you purchase a cross-cut shredder and not a strip-shredder. It is relatively easy for a thief to piece together the strips created by a strip-shredder. It is much more difficult, if not impossible outside of a forensics lab, to recreate a document that has gone through a cross-cut shredder.

Safety when entering passwords

Dave was in a casino in Las Vegas. He sat down in one of the restaurants, opened his laptop and logged into his bank to do some financial transactions. He didn't notice that a woman behind him to the left was watching as he logged in. She wrote down his username and password. Later that night, after Dave was sleeping, she went into the hotel business center and logged into his bank account. Within minutes she ordered a wire transfer of half his account balance. By the time Dave noticed the transfer had cleared and he lost the money.

Be observant when you enter your passwords when you are in a public place. Make sure no one is looking over your shoulder or can observe your password or PIN entry; this is called **shoulder surfing**. Be aware of what is going on around you and the location of people as you login to your accounts. Ask yourself if anyone can see your hands. If it appears your hands are visible, you should either move or put something in the line-of-sight to block their view.

Best Practice
Make sure no one can see you enter your username and password. Check for people around you and security cameras.

Take a few minutes to ensure that your keyboard cannot be observed by someone outside. It is not that difficult for a person in an apartment across the way to use a binoculars or even the zoom on a camera to see what you are typing.

As a side note, when you enter your PIN on an ATM machine or debit card swipe, be sure to use your hand to block anyone from watching you enter the PIN. For example, it is very simple for someone in line at the checkout in the grocery store to observe your entry of these numbers.

Not as obvious is that you should also make sure your keypad or keyboard is not in the line of sight of any security cameras. The entry of your password can be caught on those cameras and will be preserved to be potentially looked at later.

For example, in casinos not only are there literally hundreds of security cameras on the ceiling recording everything, but there are also dozens and or even hundreds of people hidden around the place, behind mirrors, in the ceiling, and walking the floor. Not all of those people are nice or have good intentions.

Maintain your privacy

By far the primary danger to your privacy is not Facebook, the NSA, or big corporations. Not by a long shot. A major reason why your privacy is in danger is *you*.

Of course there are dangers to your privacy that are out of your control. You definitely are not to blame when a big box store gets hacked and tens of millions of customer records are stolen. Naturally you are not responsible for the massive privacy issues of faulty software and bad security all over the world.

There are, however, vast amounts of data posted on Facebook, Pinterest®, Google+, and tens of thousands of other sites all over the web. There are tens of millions of blogs, millions of vlogs (video blogs), millions of web sites, and countless other places where information is posted.

Best Practice

Don't post anything on the web that you wouldn't put on the bulletin board at work, home, or the grocery store.

Individuals spend countless hours updating all of these web sites with data about themselves, their families, their work, and their friends. Billions of pages of information is freely and gladly made available to the entire world by people in the course of their daily activities.

My mind is boggled by the amount of intimate details people post about their lives, their medical histories, sexual habits, political beliefs, and everything else under the sun. I recently saw a series of posts on Facebook where people were publicly chatting about their mental disorders. I had to shake my head in disbelief. Why on Earth would anyone be writing in a public forum about this kind of thing?

Something you need to keep in mind is not everyone who looks at the information you have made public has your best interests at heart. Those words you wrote can be seen by anyone, in many cases, and can be used against you.

- These days it is normal for a company to search the Internet before making a hiring decision. They will examine everything they can find out about you, and *this is totally legal since it is publicly available information.*
- Insurance agencies may look at your public information before deciding to issue you a policy.
- Are you applying to a prestigious school? Chances are they will examine your web presence before making a decision.
- Thieves can scan through the information you make publicly available to find out where you live, where you work, what kind of valuables you own, and even your vacation schedule.

I could go on and on, but the point is that the information you post online in public forums is available for anyone to use for any reason.

Social media sites such as Facebook have many privacy settings, and you should use those to tune the information that can be seen by the public. However, the best assumption for these kinds of sites is that everything you post can be seen by anyone.

Don't blame Facebook or other social media sites for their perceived lack of privacy. You can prevent the entire issue by controlling the information you post.

Special concerns of hotel and other public computers

Sally was staying at a very nice, five-star hotel. She took the elevator down to the business center and checked her email on one of their three public computers. Unfortunately, the computer had been hacked and her username and password were stolen. Because Sally gave all of her accounts the same password, the hacker was able to log into her bank account. When Sally got to the airport she was amazed that her debit card kept getting declined for lack of funds. The hacker had wired her entire balance to a bank in Nigeria, leaving her penniless for a few days until the bank completed its fraud investigation.

I know the story. You've been traveling and you are staying in a hotel for the night. You're in a hurry and need to check your email, log into Google docs, or whatever, and you just don't want to take the time to set up your laptop. Sometimes you just to print a spreadsheet for your boss, or you didn't bring a computer at all and need to check your email.

You run downstairs and find the hotel has a business center with a couple of computers and a printer. You breathe a sigh of relief. Now you can get your work done. In fact, since you're going to be on the computer anyway you may as well log into your bank to transfer some money around, check your email, and log into your credit card account to check your balance.

Don't do it! Really, don't use the computers in any hotel business center for anything sensitive or private. In fact, don't use any public computers to access your private information.

Best Practice

Do not use public computers to access any of your personal or private data under any circumstances.

The only thing I use a hotel computer for is to print out my airline boarding pass, and that's only because I don't need to log into the airline account to do so. I just enter my confirmation number and name and then print the pass. Since my account is not accessed, it's pretty safe.

Public computers are everywhere these days.

- Hotels.
- Libraries.
- Schools.

- Parks.
- Recreation areas.
- Senior centers.
- Goodwill Career Centers.

Unfortunately, public computers are incredibly easy to hack. A malicious individual can simply insert a USB key or flash drive containing their hacking software and infect the computer in a matter of minutes.

A criminal will typically install **key loggers** onto public computers. A key logger is an application designed to record every keystroke that is typed. No real technical skill is needed to install these applications. Key logging applications are available in the **hacker underground** for a few dollars. Sometimes these malicious applications will remain undetected on public computers for months or even years, sending usernames and passwords of thousands of people to the criminal to use for his or her own purposes.

Here are some of the risks associated with using public computers

- They are often installed by people who have not been trained or are unaware of basic security principles and are usually poorly maintained, or not maintained at all.
- Whoever installs the computer may create accounts with weak passwords and elevated privileges.

- Nefarious public computer users have plenty of opportunity to hack the computer.
- Public computers are often the cheapest systems that can be found, and they generally run older, vulnerable versions of Windows. This leaves them very vulnerable to security breaches.
- They may not even have any antivirus or firewall products installed at all, and if they do these products will generally not be updated regularly.
- They are often not patched, leaving them very vulnerable to security problems.
- Visitors may browse any web sites they desire and, since the computer does not belong to them, will have no qualms about visiting questionable sites, thereby creating an avenue for introducing malware.
- Visitors may download and install unauthorized (and potentially unsafe) programs or applications.

Concerns about using other computers

The problem with using any computer other than your own is you generally don't control the security of that system. Public computers virtually always have exceedingly poor security and should never be used to access any personal information of any kind.

The same is true for computers belonging to your neighbor, your relatives, your best friend, your worst enemy, or even the person whose computer you are fixing as a favor. Generally you have little to no understanding of how well these people manage the security on their systems.

Many people do not install any antivirus program at all. Some even turn off the standard Microsoft security programs because their "system is running too slow". They may be looking at pornographic web sites (a sure fire way for your computer to get infected by malware) or downloading so-called free (hacked) versions of applications (known as Warez.)

> **Best Practice**
> *Do not access your private or personal data on any computer on which the security has not been verified.*

How do you get around this? Bring your own laptop with you. Never use a computer of unverified security to access any personal information. Sure, browsing the web is fine, but don't go beyond that. Don't check your email, your bank, your credit card balances, or anything else. Use your own computer for that.

Letting other people use your computer

It's almost inevitable when someone visits your home that they are going to ask to use your computer. They may just need to check email, log into their bank account, play some games, or catch up on work.

Use caution when allowing others to use your computer. Many people can be quite blasé about computer security either through lack of knowledge or simply not caring. Remember it is your computer and if someone asks to use it you have the right to say no.

Ideally, keeping a second computer available for guests is a great idea. Many of us have older computers that became obsolete or slow and were replaced by a newer model. Instead of throwing those systems out, reformat the hard drive, reinstall the operating system, and keep it in the closet as a loaner system. That way when a friend needs to search the web or get something done on the Internet you can turn it over them without worry.

> **Best Practice**
> *Use caution when allowing others to use your personal computer. If you must them to use it, segregate their access with a guest or their own Windows login.*

Obviously, not everyone cannot afford to have a second computer available just for guests to use when they visit. If you must allow others to use your computer, make sure you create a guest account with very restrictive security settings. At the very least, if you haven't had time to set up a guest windows account, spend few minutes and create a Google Chrome user for them. This at least isolates their cookies and surfing history from yours.

> **Best Practice**
> *Set up a guest network to allow guests to browse the internet.*

Sometimes guests will need to connect to your network. Even this has dangers, as any computer system or device on your network exposes all the other systems to some risks. To safely allow others to connect to the Internet from your home, you can create a guest network, but you will need to install your own router to do this or get instructions from your Internet provider.

If you have several people living in the same household who use the same computer, give each of them separate Windows accounts.

Never share usernames and password with anyone. Obviously there are exceptions for things like joint bank accounts and other accounts shared by spouses. Except when absolutely necessary do not give out your passwords. If there is some emergency reason why you need to share a password, change it as soon as possible.

> **Best Practice**
> *Never share passwords with anyone unless absolutely necessary.*

Your computer is valuable and so are your credentials. In most cases the data on your computer, on the hard drives, is far more important. It is in your best interests to treat your computer as you would any other very valuable thing you own.

Protecting your network

The DSL or cable modem supplied by your ISP

When you sign up for Internet service the DSL, FIOS, or cable modem that is installed by your provider may or may not come with a firewall. It is safest to assume that it does not; thus you should add some additional protection.

Here are some of the problems with the vendor supplied modem:

- It is not under your control; it is under your Internet provider's control. Since this is the gateway into your home network (and everything attached to it) it is best to isolate your network from the Internet provider.
- The equipment may be poorly secured with a weak or nonexistent password, poor security settings, or improperly secured wireless.
- The equipment may not be properly updated with new versions of the router software, which means it may have security vulnerabilities that have already been fixed.
- Anyone from your Internet provider can gain access to your router and thus can potentially get to your computers and network.

The Internet provider usually supplies a modem which typically includes four wired network connections and a wireless router. The technician who comes out to install the modem will set up the wireless for you. Unfortunately, these technicians are not usually trained in security so the wireless network may allow intruders to gain access to your network.

You also will not generally be given privileged access to the vendor router. This means you can't change security settings yourself, so you are pretty much stuck with what you are given. You could get on the line with their technical support department to make it more secure, but who wants to do that?

So what do you do to protect yourself? I recommend you purchase an inexpensive router, generally less than a hundred dollars, and install it between the one given to you by your Internet provider and your network.

Using your own router

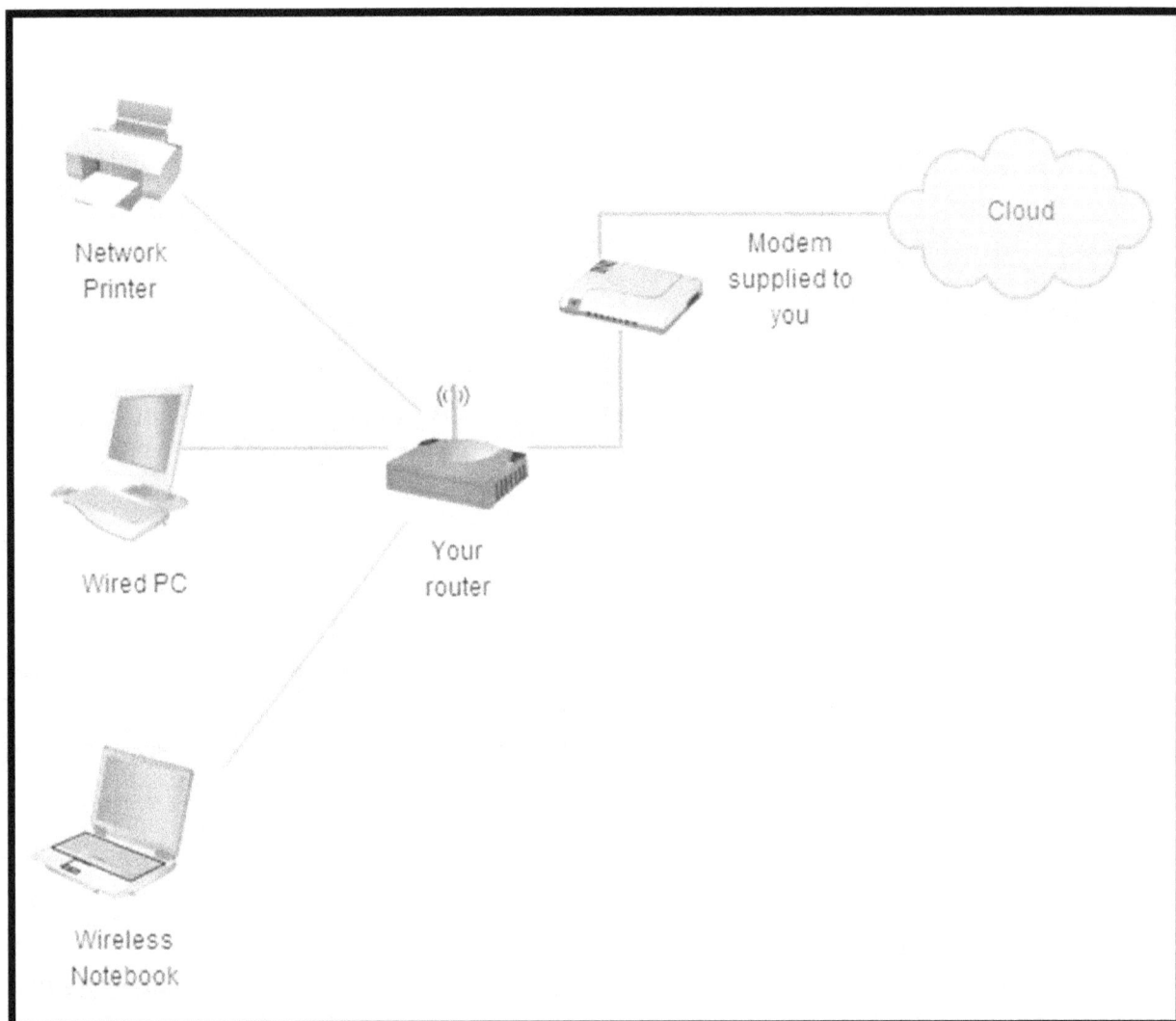

Network Printer

Cloud

Modem supplied to you

Wired PC

Your router

Wireless Notebook

One good way to keep your computer safe from malicious people is to purchase your own router. These are generally not very expensive. In fact, the NETGEAR N300 Wi-Fi Router, model WNR3500Lv2 is less than $50. This router, or one like it, has some very good features which will make your life easier.

- Very fast wireless speeds (300 Mbps).
- 4 wired LAN (internal network) connections to connect directly to your computer, printers etc.
- 1 wired WAN (connected directly to your cable, DSL or FIOS modem) network connection.

- Very simple to set up.
- Excellent security options.
- You can configure a guest network.

If you can spend a little bit more money, check out the routers on this page:

http://goo.gl/9IhfT7

http://www.pcmag.com/article2/0,2817,2398080,00.asp

Any of the routers listed will work very well in a home environment.

Once you have purchased your router, generally speaking, perform the following steps. Note this is not complicated, but you may want to have these steps performed by your favorite technical person.

1. Unpack your new router.
2. Turn off the modem supposed by your Internet provider.
3. Shut down your PC.
4. Find the cable that goes from the Internet provider's modem to the back of your PC and unplug it from the PC.
5. Plug that cable into the port marked WAN on the router that you purchased.
6. Your router should have a network cable in the box. Plug one end of that into the same port on the back of your PC where the modem was plugged in before.
7. Plug the other end into one of the other network ports on your new router. Most routers have 4 ports (these are called LAN ports) but some will only have one or two.
8. Make sure you've plugged in the power to your new router.
9. Turn the Internet provider modem on.
10. Turn your new router on.
11. Reboot your PC.

At this point your computer should be able to connect to the Internet without problems.

The manual that comes with your new router should contain the following information.

- How to connect to the router. Generally you enter an IP address such as http://192.168.100.1 (this will most likely be different for your router) into your web browser (Chrome, for example.) This is called the **router IP address**.

- The username and password needed to log into the router.

Once you have properly installed your router and your computer can connect to the Internet, perform the following steps. See the manual provided with your router for the exact instructions.

1. Connect to the router using your web browser. Enter the router IP address just like you would any other web site address.
2. The router should display a login screen. Enter the username and password you found in the documentation.
3. Change the password of your router following the procedure in your manual. Make sure you record this password in a safe place. If you lose it the only way to login to your router is to perform a factory reset.
4. Update the routers firmware. The documentation will describe how to do this. Usually there is a button or link that you click.

Best Practice

Change the password of your router immediately.

There are many, many more features that you can configure to make your network faster and better (or worse if set incorrectly). These will be described in your router documentation.

Setting up the wireless network is a little more complicated and is fully described in the section titled *Protecting wireless*. That section uses the NETGEAR N300 Wi-Fi Router, model WNR3500Lv2 as an example of setting up the security on the wireless.

Protecting Wireless

Advanced Security Feature

Do you have a wireless network? Did you realize that nine times out of ten these networks are set up with poor security that will allow any hacker to use it anytime they want?

Most modems installed by an Internet provider include a wireless network. As discussed in the previous chapter, I recommend you install your own router that includes a wireless network. This gives you control of the security of your network. If you don't install your own router, you are more or less stuck with the wireless security provided by your Internet vendor.

Setting up wireless, especially troubleshooting, can be a complex process. There are many things that can go wrong, including the following:

- The wireless is set up incorrectly.
- You have devices which do not work with your wireless router.
- Interference from other devices such as microwave ovens, baby monitors, cordless phones, and so on.
- Your neighbors' wireless may be interfering with yours.
- Your computers and devices may be too far from the router to get a signal.

If you are not familiar with setting up and troubleshooting wireless, it may be best to ask your local computer specialist to help. You can use this chapter as a guide for the proper security settings.

Best Practice
Never share the passphrase of your home network with anyone. Define a guest network to allow others to have access.

Setting up your home wireless network

Wireless Settings

Wireless Network

☑ Enable SSID Broadcast

☐ Enable Wireless Isolation

Name (SSID):   MyWireless

Region: North America ▾

Channel: Auto ▾

Mode: Up to 300 Mbps ▾

Security Options

○ None

○ WEP

○ WPA-PSK [TKIP]

⦿ WPA2-PSK [AES]

○ WPA-PSK [TKIP] + WPA2-PSK [AES]

Passphrase: MyWireLessPassphrase   (8-63 characters or 64 hex digits)

Apply   Cancel

Enter the Router IP address into your browser to access your router. Enter the username and password. One of the options will allow you to change your wireless settings. Click on the link to bring up that options screen. An example, and yours will probably look different, is shown above.

Best Practice
Do not use WEP security. Hackers cracked this a long time ago.

The setting you need will be called something like "Security Options" as you can see in the example. The setting you want is WPA2-PSK (AES). This will provide the maximum wireless security.

You will also need to set a Passphrase, which is the password for your wireless network. Use something complex such as My@Sister#!#Great. All devices that connect to your wireless network must specify the exact same passphrase. This includes any computers, laptops, tablets, cell phones, games consoles, cameras, printers, or anything else that needs to use your wireless network.

Setting up a guest network

If you are anything like most people, you will occasionally have guests over to your place. It is quite common for guests to need to connect their phone, laptop, or tablet to the Internet while they are in your home.

Avoid giving anyone your network passphase. Instead, assuming you've installed a reasonably good router of your own, create a guest network. This will allow your guests to access the Internet but NOT your own home network. If you have not purchased your own router yet, ensure it has that feature before spending the money.

When you connect to your wireless router using the Router IP address, look for the section about setting up your guest network.

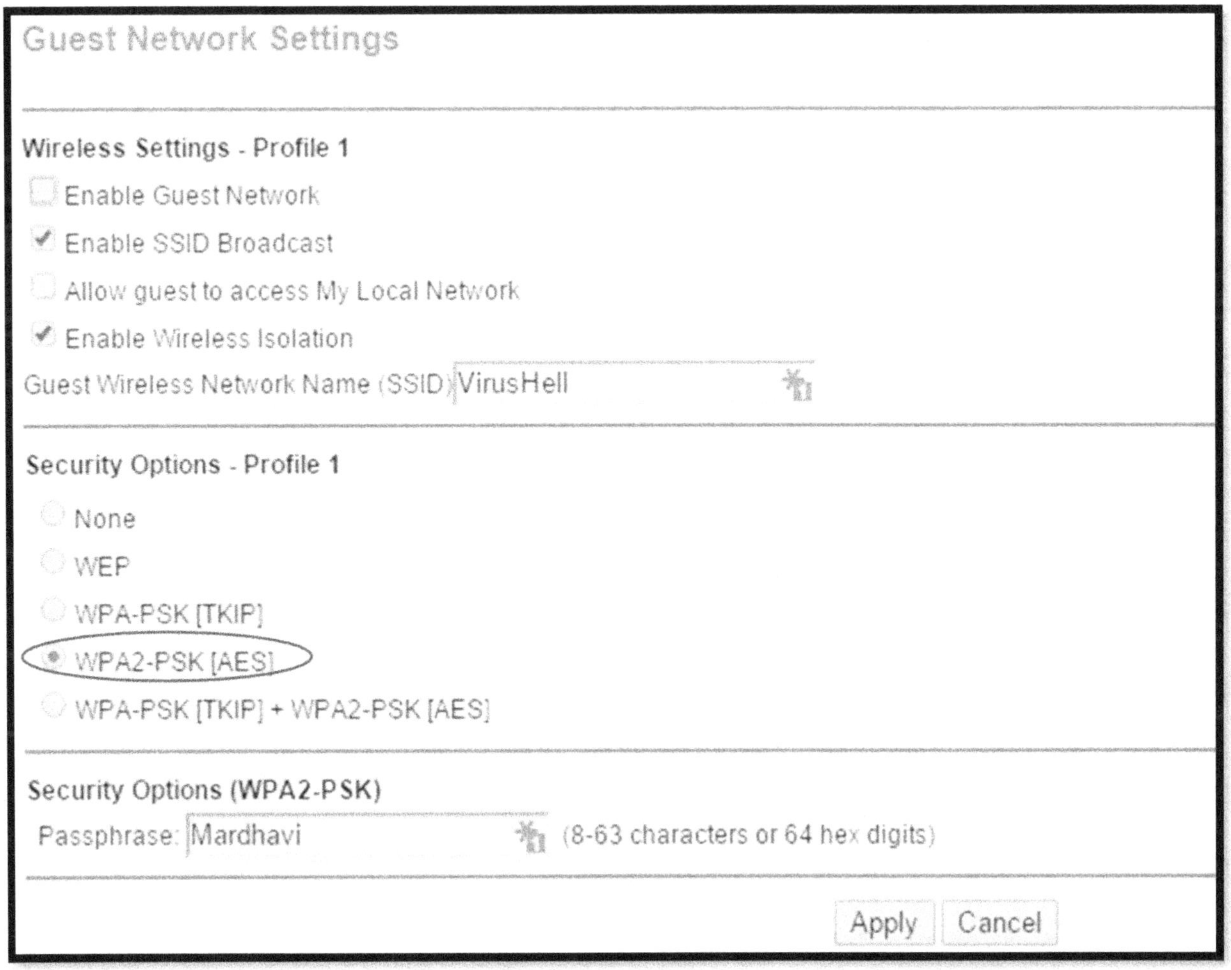

Guest Network Settings

Wireless Settings - Profile 1
☐ Enable Guest Network
☑ Enable SSID Broadcast
☐ Allow guest to access My Local Network
☑ Enable Wireless Isolation
Guest Wireless Network Name (SSID) VirusHell

Security Options - Profile 1
○ None
○ WEP
○ WPA-PSK [TKIP]
● WPA2-PSK [AES]
○ WPA-PSK [TKIP] + WPA2-PSK [AES]

Security Options (WPA2-PSK)
Passphrase: Mardhavi   (8-63 characters or 64 hex digits)

Apply   Cancel

Set the following in that screen:

- *Enable Guest Network* should obviously be turned on
- *Enable SSID Broadcast* allows your guests to see the name of your guest network.
- *Enable Wireless Isolation* should be checked. It means your guests can access the Internet but not your home network.
- Give your network a name (called an *SSID*). Your guests will look for this in their wireless settings on their device.
- Set the security to WPA2-PSK (AES.)
- Set a Passphrase (password) for your guests to use to login to your guest network. Do not make this passphrase in any way similar to your own wireless passphrase.

Now your guests can connect to your wireless network and access the Internet. This will still give you good security on the guest network and prevent unauthorized people from using your Internet connection.

Be sure to change the passphrase every few months.

Physical security

Sandra left her tablet sitting on the table "just for a minute" while she ran to get her order at the local fast food place. When she returned the tablet was gone. She searched frantically but it was nowhere to be found. A few days later she found her accounts had been hacked. It took her several weeks of effort and a lot of stress before she recovered.

L aptops and tablets are wonderful tools. They are portable, light, and extremely powerful.

One of the worst possible scenarios you can face is the theft of your computer system. All of your files and personal data stored on the computer will be not only gone, but potentially they can be viewed, erased, sold, or used by the thief. Unfortunately, the damage doesn't stop there.

Quite often laptops and computers are stolen for quick resale for a fast buck. They could be sold on Craigslist®, eBay, at a pawn shop or even at a local swap meet.

Of course, there is always the possibility that your equipment was stolen specifically to get information from you. A paparazzi might, for example, steal a laptop of a famous actress to attempt to get scandalous photos, or a business rival might

Best Practice
Do not store confidential or highly personal data on a tablet or laptop which will be used in a public place.

want information about your new product. Mobile computers such as tablets and smartphones are especially vulnerable to **snatch and grab** attacks.

Remember that your computer disk contains not just the operating system and applications, but also all of your data. This includes your photos, spreadsheets, documents, presentations, and everything else you've ever done on that system. Even if you store all your files in the cloud, local copies may be kept on your computer to enable you to work if the Internet is not available.

All of this is exposed not only to the thief, but to whoever inherits the computer system or its disks afterwards. It is quite common to find disks or computers on eBay which contain personal photos, business files, and confidential (or even secret) data.

Out and about

Years ago one of our senior managers was going around the town with his laptop. He left it on the seat of his car while he went into the store to make a few purchases. When he returned the window was smashed and the laptop was missing. There was a huge panic around the company as the laptop contained some very sensitive financial spreadsheets. A few days later he received a call from the thief demanding a $500 ransom. The manager agreed but the thief got spooked and was never heard from again.

As you go about town with your laptop, remember to practice good security. Your laptop is small and light and easy to hide. Someone could grab it if you leave it unattended even for a few minutes. Imagine the panic and horror you would feel if you left your laptop on the table at your favorite coffee shop while you went to the restroom, only to return and find it missing.

Best Practice
Never leave your computer unattended in a public or unsecured location.

Look at it this way: you wouldn't leave your wallet or your purse sitting on a table without someone watching it, would you? Your laptop is worth money and would be a hassle to replace, and the loss of the data might be catastrophic.

As the story above illustrates, it is possible for a thief to ask for a ransom for your computer. Do not under any conditions agree to arrange a meeting with the criminal. There is no telling how much harm he may cause you. He could rob you, mug you, or hold you for ransom. Instead, call the police and let them handle it.

When a criminal sees a laptop or a tablet in your car, it is like giving him a free gift. You've left several hundred or thousand dollars sitting in plain view begging to be stolen. The thief will smash your window and grab your electronics in a quick mi-

> **Best Practice**
> *Do not leave your laptop or tablet in a visible location in your car while you are away.*

nute and be gone before you know it. Now you will be missing your computer *and* you will have a smashed window to repair. In the event that you must leave your laptop in a car, put it in the trunk, out of plain site.

Make sure the room where your laptop or tablet is stored can be physically secured with a lock on all openings (doors and windows.) This includes your home or apartment, your office, and anywhere else where

> **Best Practice**
> *Lock the doors and windows of the room where your laptop is stored.*

your computer might be located. In a hotel, make sure you store your computer in the room safe or the hotel safe deposit box. Why make it easy for a criminal to steal your computer?

When you are carrying your computer with you it is a good idea to use a briefcase, carrying case, or a bag rather than a laptop case. This is especially true when transporting your system to airports, bus stops, train stations, or other highly public areas. A laptop bag tells everyone you have a high value item for them to steal.

LoJack

One of the first things a thief will steal if he breaks into your home is your computer system, especially laptops and tablets. This kind of equipment is high profit and very easy to sell. Even older computers can be sold on eBay or in pawn shops and make enough money to pay for someone's drugs or other addictions.

Best Practice

Install LoJack or a similar tracking product on your computer system.

Stolen computer systems are very difficult for the police to find and recover. However, you can make their job a little easier by installing LoJack for Laptops®. This software works exactly like the famous version for automobiles. It allows you (and the police) to find out where in the world your computer is located.

Purchase the software (it's about $40 a year per computer) and install it on your computer system. During the installation process a small locator application will be installed. Every few hours this application sends the computer's current location to the LoJack server. You can log into the LoJack site anytime and find out exactly where your computer resides.

http://lojack.absolute.com/

Most of the laptops sold these days (excluding the really cheap systems and Netbooks) have LoJack built into the hardware. This means that once you purchase, download, and install the LoJack application it cannot be removed. The thief can even format the hard drive and LoJack will still protect your system.

Of course, if your computer is stolen you should call the police. Do not attempt to recover the computer yourself. Thieves are generally not forgiving when someone invades their territory and you could get hurt.

Logout or lock your screen

Sally put the finishing touches on her bank account, then walked away from her computer screen to go to the store. Her teen age daughter was home with friends. Her daughter was watching television when one of her friends wandered into Sally's home office, saw the computer and started downloading and playing various games. Within minutes Sally's computer was infected by a dozen viruses, and it cost Sally several hundred dollars to hire an expert to remove them.

When you are going to leave your computer for any length of time, ensure the screen is locked. The best way to do this is to log out, which will then require a password to log back into the computer.

You never know who could enter your home when you are not around. If you live in an apartment, the landlord can legally enter at any time. Emergency personnel may, of course, enter if the need arises. Friends with keys or unexpected guests can show up.

Best Practice
Always log out of your computer when you leave your home or work.

By ensuring you are logged out whenever you leave your premises you will eliminate the possibility of someone easily getting into your computer simply by sitting in front of it.

Your work computer

I was managing the desktop computer systems of a large company a few years ago. Laptops were sent to my department to be repaired. I remember one instance where we found a massive amount of pornography on the hard drive of a manager's computer. Of course, my team reported it to the HR department and I am sure the manager had a very embarrassing (to say the least) conversation with them about his personal use of work equipment. It was quite common for us to find pornography on work computers, including workstations that were only used at the office.

It is common for companies to give employees personal computers, tablets, and phones to take home and on the road for business purposes. A manager could use the equipment to do reviews and other work from home, while a traveling salesman might use an office tablet to take photos of products.

Remember that this equipment is the property of the company. This means they can take it back from you at any time without notice and do whatever they want with the system. They can format the hard drive and give it to someone else and they certainly have the right to examine the computer and its contents. In fact, it is quite likely that your Internet browsing history and email messages (whether personal or business) will be available for them to see.

> **Best Practice**
> *Do not use company equipment for anything personal. This includes computers, tablets and smartphones.*

I read an article in the newspaper which told of a police officer who had been having a torrid affair with another officer. He was using a work phone for texting this lady. When the police did an internal investigation they found more than 6,000 text messages on the phone, which clearly proved not only the affair, but that it had been going on during work hours. The lesson here is that since the equipment, in this case a phone, belongs to a company and not you, they can confiscate it and give it to investigators without your permission and without giving you any notice.

Universal Security Slot

Virtually all computers have what's called a **Universal Security Slot** built into the chassis. This is a small slit, about half an inch long (half the size of a USB port) located on the side or back of your laptop. The slot often has a "lock" symbol next to it.

> **Best Practice**
> *Use a security cable to tie down your computer in public places to prevent casual theft.*

You can purchase a special cable with a lock that fits into this slot to secure your system. These cables cost anywhere from $5 to $20 each and have one tab which fits right into the slot.

The idea is to wrap the cable around something solid and immovable, such as a table leg, and slip the tab of the cable into the slot on your computer. You turn the key or set the combination (and pull out the key if that's what you used to lock it) and your computer is now secured to the furniture.

These cables provide some security, but it's best not to depend upon them very much. They will prevent a casual thief from simply grabbing your computer, but they are not that difficult to circumvent. A good pair of bolt or wire cutters is all that is needed to cut the cable.

Protect your power

Thirty years ago I worked for a computer consulting company. I was on-site late at night, past midnight, and a huge storm was going on at the time. Suddenly a bright light flashed, blinding me for a few moments, and there was a tremendous clap of thunder. It was so loud my ears rang. Everything went dark and the computers stopped working. The internal electronics of all of them were fried. The smell was horrible.

You may not realize it, but your computer is vulnerable every second of every day to power surges, spikes, sags, and outright power failures. If your system is plugged into the wall, it can be harmed. Additionally, your computer can also be damaged due to power surges via the network or phone cable.

A surge is a sudden very large burst of power caused by lightning or a piece of equipment such as an air conditioner. These can cause a computer to crash and may even destroy it completely. A sag is exactly what it sounds like, a momentary reduction of power. These are hard on computers in that they cause electronics to degrade faster than normal. A spike is simply a quick jolt of power, also hard on electronics.

> **Best Practice**
>
> *Plug your computer and your computer accessories into a UPS to protect it against power surges and failures.*

How do you prevent problems from these various electrical issues? The solution is simple and cost effective. Purchase an **Uninterruptable Power Supply** (UPS) and plug your computer into that instead of directly into the wall. This device will provide power to your computer for a few minutes if the power fails, and can be configured to shut down your computer automatically if needed. If you are on a budget, at least use a surge protector instead of plugging your equipment directly into the wall.

All UPS devices include surge protection, so they will protect you from power sags, surges, and spikes. You can also plug in your phone line (if you use a modem or fax line) and your network cable into the UPS to make your computer even safer.

Getting rid of your computer equipment

> *It seemed really simple. Sally bought a new computer. She didn't need her old system so she sold it on EBay for a few hundred dollars. She had some racy pictures of herself on the hard drive but she deleted them before she sold the computer. A few months later she found those pictures on a pornographic web site. She thought her boyfriend might have sold them or something, but he claimed to know nothing about it. Most likely, someone was able to recover the photos off the hard drive of the system she sold on EBay.*

Did you know 40% or more of the disk drives sold on eBay contain personal data including financial records, photos, pornography, and even sensitive corporate trade secrets?

Operating systems (all of them) do not erase files when they are deleted. How does this work? Let's say you have a great collection of movies and you make a list of those movies. You erase one of the entries on the list. This does not erase the movie, it simply removes the line item from the list of movies.

> **Best Practice**
> *Use a product such as CCleaner to erase all of the disks on a system before you dispose of it to ensure your private date is completely gone.*

Disks work the same way. When you delete a file it is removed from the list of files on the disk. The data in the file itself is not erased; that space is made available to be reused later, but until the time it is reused, the space still contains the original data. Thus, when you delete those pictures or that spreadsheet they are actually still on the disk and can be recovered.

When you get rid of your computer, whether you sell it on EBay or Craigslist, throw it out in the trash (although that is illegal in many municipalities), or give it to a friend, your data is still there. It can all be recovered.

CCleaner is a handy application with a number of options to help you manage your computer. One of its tools completely erases (wipes) a drive so that the data cannot be recovered.

Download the free version of CCleaner from the following web site. Note the professional version has some useful features which might make your computer run faster or better. The free version includes the disk wipe so you don't need to pay just to use that one feature.

https://www.piriform.com/ccleaner

Install the program and run it. Click on the "Tools" button on the left side, then click the "Disk Wiper" button. This will display some options. Choose the drive you want to wipe (erase) and select "Erase Drive" from the "Wipe" options (circled in the picture below.)

Make sure the disk you select is really the disk you want to erase. Once the disk is erased it is gone forever. There is no possibility of recovering the data once this program has done its job.

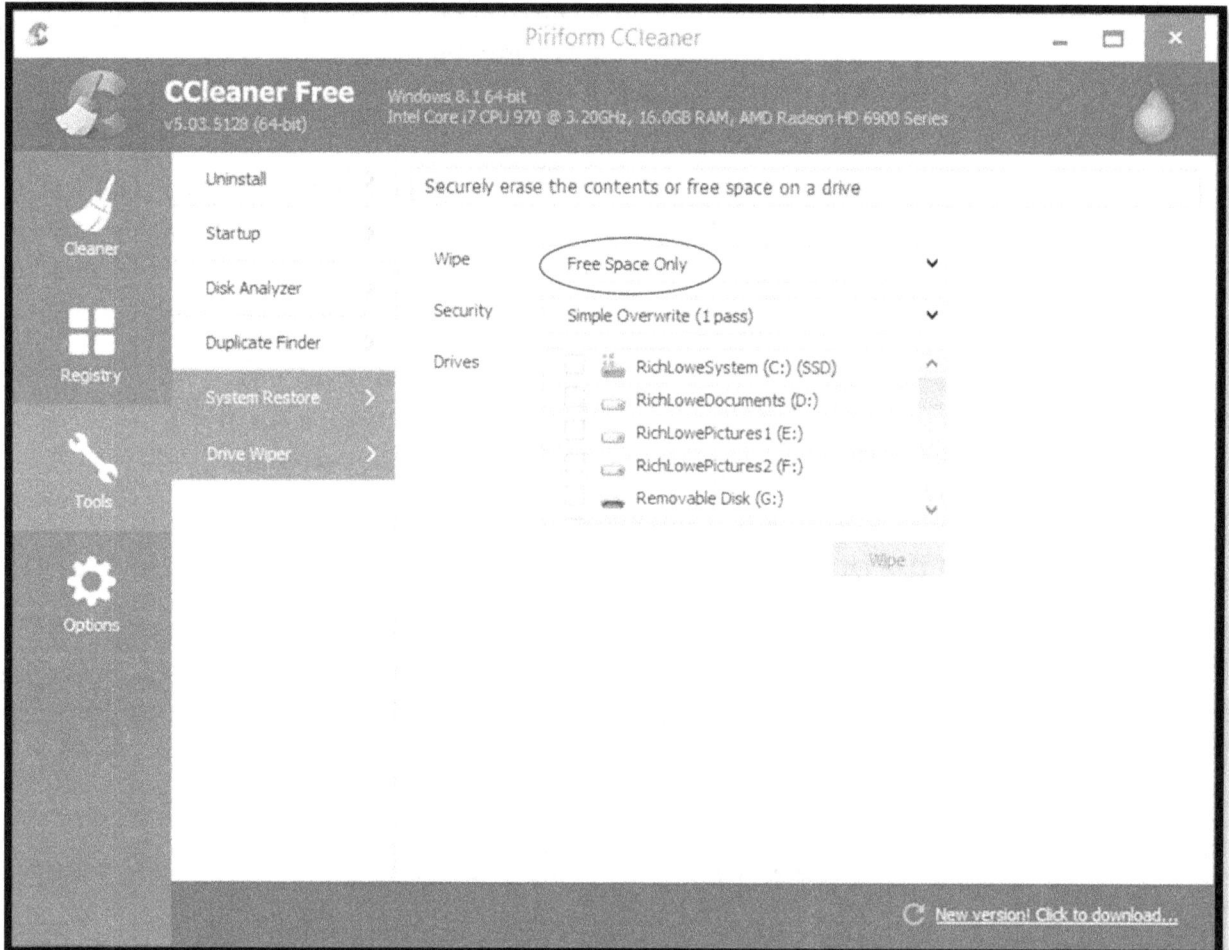

Note you cannot erase the C drive because that is the drive the operating system is using. If you want to erase the C drive you need to either:

a) Remove the drive and erase it on another system.
b) Delete all of your personal data files from the C drive, then clean it using the "Disk Wiper" in CClean. Choose the "Free Space Only" option from the selection. That will delete all of the deleted but not erased files.

Your operating system

> *David noticed his computer was running very slowly and seemed to be crashing more than normal. Upon investigation, he found his computer was infected by a dozen serious viruses. He thought Windows Update was keeping his system current with security fixes. The technician who looked over his computer informed him that he was running Windows XP, an obsolete version of Windows which is no longer supported by Microsoft. This meant Windows Update was no longer working. Thus over time his system became more and more vulnerable to viruses and other malware.*

Your computer might seem simple. After all, you just plug it in, push a button and in a few seconds it comes up and you do whatever you are going to do with it. But underneath the covers it is actually a very complex device. Your computer consists of **hardware**, which is the machine and all its parts, and **software**, which is the set of instructions which make the hardware operate.

This software, called the **operating system**, is what makes the difference between cold metal and plastic that does nothing and a machine on which you can play games, write novels, and surf the web.

The most popular operating system in the world for personal computers (workstations and laptops) is Windows. Over the last few decades, there have been several different Windows operating systems, beginning from Windows 1.01, which was released in November 1985, up through Windows 8.1, released in October 2013. Windows 10, the newest version, is due to be released in July of 2015. Each of these different flavors of Windows is called a **version**.

Each of these Windows versions served their purpose when they were released. They work great (and sometimes, as in the case of Windows ME, not so great) for a few years, then newer and better computers are invented. The old Windows versions become obsolete and eventually Microsoft no longer supports them.

The problem with these older Windows versions is there is a war going on between hackers and computer manufacturers. Hackers are constantly coming up with new ways to break into computers, and the manufacturers are finding ways to keep them out. As operating systems age beyond a decade, the manufacturers no longer keep them up-to-date with security fixes.

This means the older computers tend to be much more vulnerable to hacking than the newer systems. Windows 95, for instance, has not been supported in years, and thus is extraordinarily vulnerable to hacking. Windows 8.1, on the other hand, tends to be very secure as long as you ensure security updates are regularly installed.

People use older computers for many reasons. Sometimes they cannot afford to purchase a newer model of computer or they don't want to take the time to learn the newer version. Commonly the older computer does what is needed very well so there appears to be no particular reason to go through the hassle of upgrading. Occasionally there are applications which will not work on the newer operating system, thus making it painful to upgrade.

> **Best Practice**
> *Upgrade your operating system to the newer version within a year of its release.*

Generally it is a good idea to upgrade your operating system within a year or two after the newer version is released. I like to wait a year or so after the initial release as this allows some time for many of the inevitable bugs to be found and corrected by the vendor. Quite often the first release of a new version (sometimes referred to as the ".0" version) of an operating system can have quite a few problems which need to be resolved.

To help with the upgrade process, vendors often provide a tool to examine your devices and applications to ensure they are compatible with the newer version. This tool will give you instructions on anything it finds which needs to be handled to ensure the upgrade is successful.

As you can see from the list to the right, the only secure Windows operating systems are Windows 7, Windows 8.1 and Windows 10. Note that Windows 8 is obsolete and should be upgraded to Windows 8.1 immediately.

Windows Operating Systems
➢ Windows 3.51 – Highly insecure
➢ Windows 95 – Highly insecure
➢ Windows 98 – Highly insecure
➢ Windows ME – Highly insecure
➢ Windows 2000 – Highly insecure
➢ Windows XP – Insecure
➢ Windows Vista – Highly insecure
➢ Windows 7 – Reasonably secure
➢ Windows 8.1 – Extremely secure
➢ Windows 10 – Very secure

What should you do if your computer is running one of these older versions of Windows? Either upgrade it to a newer version or, if that's not possible, replace the computer with a newer model.

Major upgrades, such as the one from Windows 7 to Windows 8.1, are complex and may not be successful. You should always ensure you have a full system backup before beginning any upgrade.

If you choose to get a new machine, there are several companies that sell products to help you move your files and applications over to it. I've used a product called PcMover from LapLink to perform this move and it's always worked perfectly. You can find this product at the URL below.

http://goo.gl/mol4wX

http://www.laplink.com/index.php/individuals/pcmover-for-windows-8/pcmover-home-8

PcMover costs about $40, but saves a tremendous amount of work. Download the application, install it on both computers (new and old), then run it to transfer everything from the old computer to the new one.

The cloud

I'm sure you've all heard about something called the **cloud**. No, I'm not referring to those white or grey misty shapes in the sky. For the purposes of this discussion, the cloud means resources which you use which are out on the Internet instead of being part of your network. Disk space, photo storage, and printers are examples of resources.

Why are so many people starting to use the cloud?

- More than one computer can access the same resource.
- More of the resource, such as disk space, can be added as needed. For example, if you have 5gb of space on Google Drive and if you fill that up, you can casily purchase more space.
- The hardware does not take up room in your home or office.
- You can use the same resources (disk space, photo storage, and so on) from different computers all over the world.
- Cloud services can serve as backups.
- Someone else manages those resources for you.
- Since the security of cloud services is in the hands of the cloud company (Google, Microsoft, Amazon, or whatever) you don't have to deal with it.

- You can get cloud services quickly. Since you don't require any hardware, all you do is sign up, pay the appropriate amount (if required), and start setting up and using the service. There is no wait for hardware to arrive and no hardware configuration.

There are some disadvantages of using cloud services

- Since cloud services are out on the Internet, speed can be an issue. Your performance depends on the speed of your Internet connection.
- Security can be a huge concern. You are trusting someone else with the security of your data. On the other hand, those people are generally well trained in keeping your data safe.
- Privacy may be a concern.
- Cloud services can be expensive.

I have found cloud services, when used judiciously, to be an excellent addition to the arsenal of home computing resources.

Of course, as with everything on the Internet, hacks will happen and the cloud is not immune. There will be problems, but the larger companies such as Google, Amazon, and so forth spend a huge amount money, time and resources on security as they do not want to have their Brands damaged through a security breach.

Another caution, especially with smaller services, is sometimes companies go out of business. Just make sure everything you own is also stored on your own disk drives or backed up locally.

One last note is to always, without fail, review the terms and conditions of the service thoroughly. There is a lot of legal mumbo-jumbo but you should be able to figure out the important points. If you find some parts that you don't understand either get someone knowledgeable to help you or use a good dictionary to get the definitions of the words cleared up.

Cloud file storage

The most common use for the cloud is to store files. Services such Google Drive and Microsoft OneDrive are free and can be made to look like regular disk drives on your desktop. You use them exactly like you would any other disk device on your computer. You can move or copy files to them, create and edit documents, and so on.

Some of the most popular cloud file storage services include:

- Google drive.
- Amazon Cloud Drive.
- Microsoft OneDrive.
- Apple iCloud.
- IDrive.
- Box.
- Dropbox.
- Hightail.

Your operating system can be set up to keep two copies of everything you store on these cloud drives. A copy is kept locally, on your computer, so you can still access it if the Internet is not available. Another copy is kept on the cloud drive. Anytime you modify one version, the other is automatically changed with the same information.

This allows you to share files very easily. For example, let's say you want to work on a spreadsheet from your home computer, your computer at the office, and your tablet when you are traveling. If you put the spreadsheet on a cloud drive, you can access it from any of those systems anytime you want.

Another advantage to using the cloud for file storage is it can serve as a backup. A copy of the spreadsheet (from the example above) is stored on a disk somewhere else in the country. Thus if your computer is destroyed or stolen you still have a copy of the spreadsheet safely stored in the cloud.

Cloud backup services

Backup is one of the most useful cloud services. There are several companies that will keep a copy of all of your information safely in the cloud. This provides a high level of security for your data, since it is not stored within your home or office. Your pictures, videos, documents, and other files are safe even if your computer is destroyed or stolen.

Some of the popular cloud backup services include:

- The LiveDrive service, which I use on my own computer and highly recommend, has a relatively low yearly fee, does not limit the size of backups and once installed is totally automatic. LiveDrive will back up data files on your C drive, but not system files or applications.

- Carbonite is similar to LiveDrive and is also highly recommended. Carbonite is unique in that with one of their packages they will courier your data to you in the event of disaster. If you have a very large volume of data backed up this can make your recovery much quicker and easier. Carbonite automatically backs up your data. It will back up data files on your C drive, but not system files or applications.

I run both of these services at the same time on my computer. I've done this for years without an issue. Of course this requires a fast computer and a very fast Internet service. Why do I run two backups? These are valuable files and I want to ensure that no matter what happens my data is safe.

There are dozens of other cloud backup services. Most of them are much more expensive than LiveDrive or Carbonite because they don't allow for unlimited storage.

Be aware that in order to maintain a backup of video and other large files you will need to change some settings in both LiveDrive and Carbonite. This is fully explained in their documentation.

Neither of these options makes a backup of your system (the windows folder) or your applications. Thus they cannot be used to recover the operating system in the event of a total system failure. They do, however, keep your own data safely stored on hard drives in the cloud.

As far as security goes, both of these services provide excellent protection of your data, using state-of-the-art methods.

Cloud printing

I'll bet it never occurred to you that you can print to the cloud. What this means is you add your printer to the Google cloud service. Your printer is still private and you can print to it from just about any device or computer you own.

You can print documents to one of your cloud printers from any computer, smart phone, or tablet. For example, if you had a spreadsheet open at work, you could print it directly to the cloud printer you defined at home. This is especially convenient for printing from tablets and smartphones.

To set up cloud printing, just go to this link

http://goo.gl/SyoU4u

http://www.google.com/cloudprint/learn/

Read the instructions and add your printers.

I am sure there are other cloud printing services, but this is the only one I have used.

Is this service secure? Google has a good reputation for security so it is certainly as safe as any other Google service.

Users

When you first install Windows you are asked to create a user account. You can create a Microsoft account or a local account. I recommend using a Microsoft account because settings are stored in the cloud and it automatically creates space on Microsoft OneDrive, which is cloud-based storage.

You have the option of creating a local account, which does not use cloud-based storage, nor does it store settings online.

The following web page describes how to create both local and Microsoft accounts.

http://goo.gl/QoxWtg

http://windows.microsoft.com/en-us/windows/create-user-account

A major flaw in older versions of windows, prior to Windows 8, is that accounts, by default, had administrator rights. This left the operating system wide open to attack from malware and hackers.

> **Best Practice**
> *Create a separate Windows account for each person using your system.*

Modern versions of Windows have corrected that mistake. The first account created, the one Windows asks about during the installation process, is an administrator account. User accounts that you create are, by default, not created with administrator privileges. This fixed the largest, most gaping problem with Windows security.

Every person who uses your computer should have their own Windows account. In addition, a guest account should be created for those instances when someone stops by unexpectedly and needs to use your computer.

> **Best Practice**
> *During your normal day-to-day activities, do not log into the Administrator account. Use a user account instead.*

To create accounts, select the *User Accounts* control panel. This will display a screen similar to the following:

Click the *Manage another account* link.

At the bottom of the window that is displayed, you will see a link titled *Add a new user in PC settings*. Click that to add a new user. You can create either a Microsoft account or a local account from the utility. Normally it is recommended to use Microsoft accounts.

The process for creating and managing accounts is described on this page:

http://goo.gl/Y8In4J

http://windows.microsoft.com/en-us/windows-8/microsoft-account-tutorial

Guest accounts

To allow for unexpected guests who need access to a computer, I recommend creating a guest account. You can use the built-in guest account for convenience, or, better still, you can create an account with a fake name, make it a normal user account, and change the password each time you need it.

To use the built-in guest account, bring up the User Accounts control panel as described above. Click the *Manage another account* link to display the screen shown below.

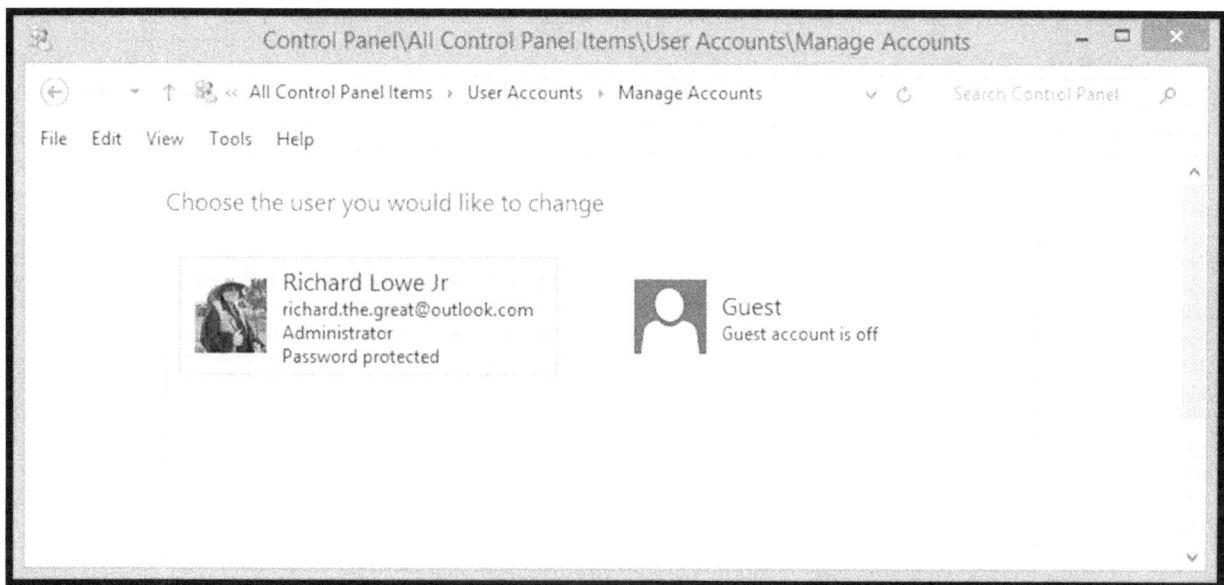

Click the *Guest* icon to display the following window.

Turn on the guest account and your visitors can log in using the username **guest** without a password.

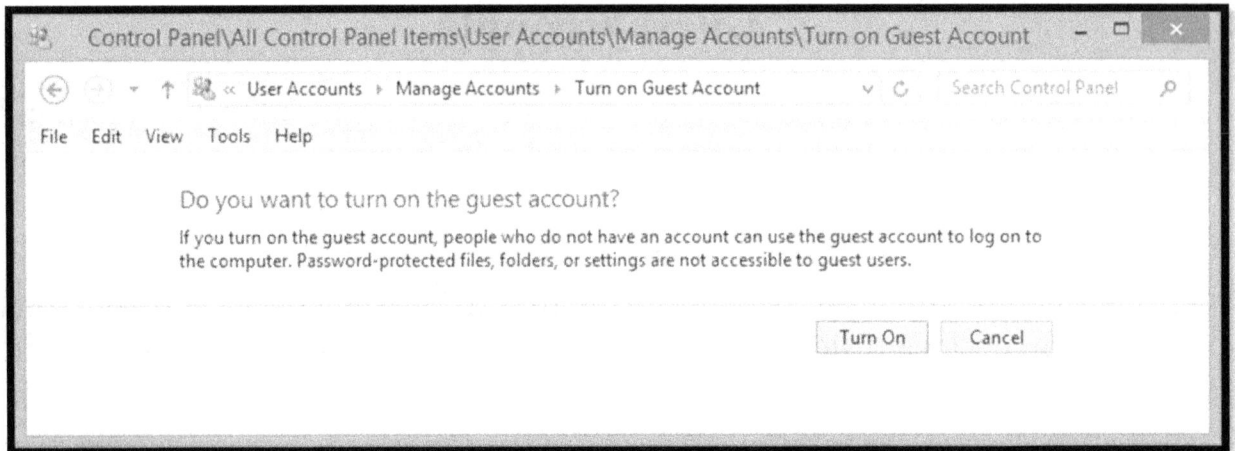

You should only enable the guest account when needed and disable as soon as it is no longer required.

Passwords and Usernames

Noah believed he was being very clever as he thought up a new password for his online bank account. He thought no one could possibly know his birthday so it seemed bulletproof; it had the advantage in that it was easy to remember as well. A month later Noah was on the phone with the bank trying to convince a very nice foreign support lady that he wasn't the one who transferred out his entire balance to a bank in Nigeria...

If you are anything like me, you've got at least a few dozen online accounts. These accounts include

- Utility accounts for water, power, sewage, gas, trash, and so forth.
- Apartment, leasing or mortgage accounts.
- One or more banks.
- Various credit cards.
- Stores where you commonly shop.
- Google, Facebook, iTunes, and potentially hundreds of other sites.

Accounts are usually set up so two pieces of information need to be entered to gain access. These are the username and the password. Often the username is your email address, although sometimes you are allowed to pick your own username.

The fact that your email address is a username on so many sites presents a real security problem. Once a hacker knows your email address he knows the username for most of your online accounts.

If you, like most people, use the same password for most or all of your accounts, a hacker who cracks one account gains access to *all* of your accounts, at least those that use your email address as the username.

> **Best Practice**
> *Use a distinctly different password for every account.*

Best practices for passwords

"**Passwords are like underwear. You shouldn't leave them out where people can see them. You should change them regularly. And you shouldn't loan them out to strangers...**"

— **Unknown**

A password is the same as a key to your information. If an unauthorized person gets access to your password they can do whatever they want with your data and your account. For example, if a hacker gains access to your online account for your electric company, he could shut off your power or cancel your service.

Here are some best practices for passwords:

- Use longer, more complex passwords that contain upper and lower case letters, numbers and, if the application or web site allows, symbols such as # and @. If possible, make your passwords at least 20 characters long.
- Do not share passwords with anyone, including your co-workers, your friends, your children, or anyone else.
- Do not write down your passwords anywhere. Not on post-it notes, not on a sticky on the bottom of your keyboard, or a piece of paper near your desk.
- Do not ever click on a link in an email which requests you to change your password, no matter how official the email appears to be. To repeat, *do not click on links in emails to change passwords or enter account information.*

Your password should *not* be any of the following:

- Any word in any dictionary.
- Your username, real name, spouse's name, child's name, any relatives name, or, for that matter, any name at all.
- A character or place from your favorite book, movie, or song.
- Abbreviations.
- Any of the above with a single special character in front, such as &tom.
- Any of the above capitalized.
- Any made up words such as xyzzy or fubar or wugga.
- An easily guessed string of numbers such as 123456.

Some examples of good passwords include:

This%iS&mYLife
ThE&LiFe$of-BRIAN%

The problem with this kind of password is they are difficult to remember and are prone to error when entered, especially on smaller devices such as phones. This is why I highly recommend using a password manager to control and store your passwords.

Password managers

The solution to this conundrum is to use a password manager to store all of your usernames and passwords in the cloud. You can think of this as a safe deposit box for your passwords, only instead of being stored in a bank

> **Best Practice**
> *Use a password manager to manager your usernames and passwords.*

they are kept in a secure location out on the Internet.

I've tried quite a few password managers and by far the best of them is LastPass. You can download LastPass from this link:

https://lastpass.com/

You only need to remember the username and password for LastPass. After that, LastPass will take care of all of the rest for you. All your usernames and passwords are safely stored online, and LastPass will fill in the blanks automatically each time you visit a web site that requires login information.

This means you can create a different password for every single account on the web that belongs to you. Each of those passwords can be very long and complex since you no longer have to remember them.

To make the process of choosing good passwords easier, LastPass has a password generation function. This creates complex, random passwords for you. Using this feature ensures your passwords are different, and difficult to crack.

LastPass also provides a feature to audit your password security and suggest where you might make changes to be more secure. For example, if it finds you have several accounts with the same password it will suggest you fix that problem, or if your passwords are too short it will inform you so you can make a correction.

A quick note about "mother's maiden name"

Virtually all web sites provide a way for you to recover or reset your password on those occasions where you forget. Sometimes you can click a link to get a new password sent to your email, and other times they ask for some information that presumably only you know to validate that you are indeed you.

These sites may ask for your mother's maiden name under the presumption that a stranger would not have this information. While this was never really very secure, in these days the massive information available on the web has made this practice extremely questionable, to say the least. Generally a hacker can find out your mother's maiden name from public records within a few minutes.

Many web sites are getting around this problem by allowing you to select from a list of questions such as "what is your favorite color?" or "what street did you live on when you were a kid?" Although this is better than your mother's maiden name it still remains vulnerable to someone researching your past.

Best Practice

Answer the "mother's maiden name" identity questions with false information.

The best way to answer these questions is to come up with some false, more or less random, answers. For example, instead of answering with your mother's maiden name, answer with some other last name. No one cares whether your answer is factually true or not as long as when the answer give matches what they have on file.

LastPass includes a place to record notes for each web site it has on file. You can use these notes to record the answers to these challenge questions. If you don't use LastPass, you'll need to record your fake answers somewhere safe so you don't forget them.

The idea is to defeat the ability of a hacker to perform some basic research and divine the answers from that.

The problem with email addresses for recovering passwords

Sally Black had an email address consisting of her first and last name, sally.black@whatever.com. When she got a divorce she made a new email account using her maiden name Sally White, sally.white@whatever.com. She thought she changed everything over to the new account but she missed the email address stored in her bank account. She let sally.black@whatever.com expire since she felt she didn't need it anymore. A few months later she found herself locked out of her bank account. After many hours on the phone trying to convince the bank that she was who she said she was, she found that all her money had been transferred out and was gone.

A hacker had recreated her old sally.black@whatever.com email account and then used the retrieve a password link to have the password emailed to that account. Once he successfully logged in the hacker changed the password and transferred out all the money.

You might think that the safest method for retrieving passwords is to get them sent to your email address. Generally this is true; email is a fairly safe way to retrieve your password.

All normal email messages are sent directly over the Internet in such a way that they can be read by anyone. In other words if someone managed to grab the email as it "went past" they could easily read the information in that email message, including the password.

This is why you should immediately change your password as soon as you can after receiving it in a message. It is possible that someone else may have read that email message and seen the password.

There is another problem with using email for password retrieval. Let's say your email address is roger@mymail.com. Suppose one day you forget to pay the annual fee for the email address or you decide you don't need it anymore. Then you forget that your bank account still uses that email address for password retrieval.

Best Practice

When a password is sent to you via email, change it immediately after you log in.

All someone needs to do is grab your old email address, then click the link to retrieve the password. Within a few seconds they will receive the password into your old email account, which they now own.

> **Best Practice**
> *Ensure your email account has the best security (password) you can.*

This problem can also occur if your email account is hacked. This is the reason why it is absolutely critical that you secure your email account exceptionally well. Regardless of any other security practices you decide to implement, always ensure your email password is very long and very strong.

Two factor authentication

There are many problems with requiring a username and a password for security. A hacker can guess them, or they can use any number of tools available on the Internet for cracking into accounts.

- Many web sites use your email address as the username. This weakens security because the hacker does not need to figure out your username. He already knows and this makes getting into your account even easier.
- If you are allowed to choose your own password then it's very likely you will choose something you can remember, which means it may be possible for a hacker to use some standard techniques to figure it out.
- Some web sites get around this by forcing all kinds of rules about how long the password must be, what kinds of characters need to be included and so on. This frustrates users and makes it more difficult to remember those passwords, which makes it more likely that they are written down on pieces of paper. The hackers can still use various tools to crack your password, albeit with a little more difficulty.

When you come right down to it, usernames and passwords are not a very good method for securing an account. It's just too easy to hack no matter what precautions are taken.

Because of this many companies are starting to use what is called **two-factor authentication**. I know this sounds complex but it is actually a very simple methodology.

Most two-factor authentication schemes are based upon these concepts:

> ***Something you have*** – *Consider your home security for a minute. Generally you have a lock on your front door and you have a key to get inside. The key is something you have. You have a key in your possession and merely having that key lets you into your home.*
>
> ***Something you know*** – *If you added an alarm system with a keypad to your home security, you now have something you know. You know the passcode to disable the alarm system. This means you need both a key (something you have) and a passcode (something you know) to get inside your home.*
>
> ***Something you are*** – *For even more security, including a fingerprint, retina, or palm scanner adds something you are. In theory this creates very strong security. Since the scanners match patterns and are electronic they can be misled, but that is very difficult.*

Your average username and password security system for web sites uses only one of these three criteria, generally *something you know*: your username and password.

Two-factor authentication adds another level of security to this scheme. In addition to a username and password, many companies require one more piece of information that you must enter, generally a code of some kind.

This is a use-once-only code created at a moment in time and sent to *something you have*, for example a cell phone (via text message) or an email account (via email message). Since the code is unique and created at that moment, it cannot be cracked. Sometimes the code is only valid for a few minutes or an hour or so. After that a new code must be created and sent to something you have.

Of course all systems have weaknesses, and the main weakness of two-factor authentication occurs when someone gains access to your email account or steals your cell phone. At that point they can get the codes and, assuming they have the username and password already, can gain access to your account.

Two-factor authentication requires something you have in your possession, which can make it a bit annoying on occasion. For example, Google sends a text message containing the secret code to your phone. If the phone has been lost, stolen, or misplaced that can be a problem. I like to keep my cell phone charging in the bedroom, so when I have to log into Google from my home office I have to get up and go get the phone.

Some laptops and cell phones use *something you are* to attempt to make security even tighter. Generally these use fingerprint scanners to validate access. Unfortunately, testing has indicated that fingerprints are relatively easy to fake, at least with the technology used on your standard cell phone or laptop. Of course the technology will get better over time and this will become more bulletproof.

> **Best Practice**
> *Use two-factor authentication whenever it is available.*

Two-factor authentication is slowly being implemented on many different web sites. Google, Facebook, and Twitter, for example, all have an option to use two-factor authentication. Many other web sites have not yet implemented this higher security, generally because it is more complex and that complexity confuses people.

Although it is true that sometimes two-factor authentication can be annoying, I've found it to be very secure. In my opinion, it is worth the added annoyances and hassle to use this additional layer of security. At the very least consider for accounts of high value, such as banking and email.

Preparing for disaster

For years I've photographed renaissance fairs, belly dance shows, and national parks, among other things. By 2008 I had more than 300,000 photos, all of them stored on a single external hard drive. Even though I've been a computer professional for years, it never occurred to me that something could happen to that brand new drive. Hard drives, at that time, were expensive and I figured I'd make a backup "tomorrow", but somehow tomorrow kept getting put off to the next day.

One day I was sitting at my desk editing a photo when I heard a horrible screech from the drive. It didn't take me long to realize that all of my photos, the originals anyway, had been destroyed. I had copies of the photos in a modified, smaller form on another drive, but the originals were gone.

With some experimentation, I found that the drive was not totally destroyed. I could turn it on and the drive would last from ten to thirty minutes before shutting down again. So over the next few weeks I spent long evenings copying files off the drive, little by little, until I recovered all but a few hundred.

With this narrow escape from disaster, I learned my lesson and quickly came up with a real backup scheme.

Use cloud-based backup

Before doing anything on your computer, check, *now*, to see if you have some kind of backup. No? Then your data is living on borrowed time. Your computer could be working perfectly one minute and in a few seconds you could have a real disaster on your hands.

Think about what is stored on your computer. Your business records? Family photos? Videos? If you are anything like most people with computers these days, you've got hundreds, thousands, or more files that are irreplaceable.

So what is the best way to keep these files safe?

I recommend a product that has literally saved me numerous times. It's called LiveDrive. Use the link below to proceed to the LiveDrive web site, download the product, install it, and begin your trial subscription.

 http://www.livedrive.com/ForHome

This wonderful little application takes a few minutes to install, and best of all, it's completely automatic. As long as your system is connected to the Internet your files will be automatically saved to a secure location outside your home. It's cheap, around $70 a year; it's simple, secure, and safe.

Best Practice
Use an automatic cloud-based backup application to store your data safely in the cloud.

But besides a catastrophic computer failure, what else could happen to your files? Well, a virus could infect your system, the hard drive could fail, your friend, child, or co-worker could delete the files, a thief could steal or wreck your computer, or a natural disaster could wipe it all out.

If you use LiveDrive, you don't have to worry about any of this. Whatever happens to your computer, your files will be safely stored so you can recover them as needed.

Another good cloud backup solution is called Carbonite. This product is similar to LiveDrive and costs about the same. Use the link below, download, and install the product on your computer. You can get a trial subscription to determine if you like the product before you purchase it.

http://www.carbonite.com/

Carbonite has an advantage over LiveDrive. If your purchase the more expensive package you can get your data sent to you on hard drive via courier in the event of a disaster. Note that you don't need to purchase the more expensive package until you need that drive sent to you. You can upgrade, paying the higher fee, at any point.

I've used both products. Each of them is excellent. They perform well, are barely noticeable when they are running, and have excellent restore capabilities. Either product will serve the purpose of backing up your system safely into the cloud.

It is important to understand, though, that certain files, most specifically videos, are not backed up unless you manually add them to your backup. Read the product documentation for what you need to do if you want to save copies of videos in your online backup.

> If your system is infected by a virus, files containing the virus can be copied to your Cloud backup (Carbonite, LiveDrive, or another product). Ensure you occasionally create backups to local media, such as CD, DVD, or disk, and save those away from your computer.

Create a local backup

A local backup is a copy of your data that you make on another disk drive or other media. This can be very convenient, and the data is at your fingertips if you need to recover something. You can create these backups on many different types of media.

- You can purchase a removable, USB hard disk drive from Amazon or just about any other computer store. These disk drives are cheap, small, and last a long time providing you don't damage them.

- Writable storage such as DVDs or Blu-rays may be used to back up your files. Many new desktop computers come preinstalled with a Blu-Ray or DVD writer (usually one device will do both). You can also get a USB DVD or Blu-Ray writer relatively inexpensively.
- If you don't have a lot of data, a SD card (the little cards used in digital cameras) or flash drive can be used. These cards or flash drives are relatively inexpensive and are available in many different sizes.

The advantage of creating a backup on a local device or media is it can be done very quickly and without a lot of preparation. All you need is the device (or media) and you are all set. To be well protected, the backups should not be stored in the same location as your computer. You might keep your home backups in a box at work or in a safe deposit box. This provides protection in case you have a disaster.

The disadvantages of using this method for creating backups of your data include

- It is an entirely manual process.
- You have to locate the folders on your computer where your files are located. If you miss any they won't be backed up.
- The media can be dropped, lost, or destroyed.
- The backups can be stolen if they are left unguarded.

The problem with local backups (disk and network drives) is if your computer is infected by a virus the local drives will also be infected. Thus you should dismount those drives, and even turn them off, when they are not in use.

Always spend an extra few minutes to check your local backups after you have created them to ensure it was successful. Just open up the media and validate that the files exist, and that the dates look correct.

The easiest method for locally backing up your files is to insert the drive or media (Blu-Ray, DVD or Flash Card) and copy your files directly to it, overwriting anything that already exists. A simple drag-and-drop of your documents folder to your removable hard drive, for example, is quick, easy, and relatively fast.

You can also use a product such as InSync to update your backups. This inexpensive product is simple to use and very fast. You can find the InSync product at the web page below.

http://goo.gl/tyWtmN

https://www.dillobits.com/detailinsync.html

This product is not difficult to use. There are full instructions on the web site.

File History

Windows 8 and above includes an automated backup system called **File History**. What this does is maintain your files and any changes you make to them on a hard drive or a network drive of your choice. File history is a great backup solution for the following reasons:

> **Best Practice**
> *Turn on file history to ensure you have a local backup of all your personal files.*

- It is fully automatic once you have set it up.
- It is a very simple-to-use solution.
- It is easy to restore your data.
- You can restore older versions of your files if you desire.
- File history is very fast.

This method does have some disadvantages

- If your computer is infected by a virus your file history backup can be infected or destroyed.
- If you have a lot of data, such as photos or videos, file history may be too large to fit on a disk drive and thus cannot be used.
- Since file history is local to your system, it does not protect you from disasters.

File history is fully described at the following web page.

http://goo.gl/nIK9xo

http://windows.microsoft.com/en-us/windows-8/how-use-file-history

Using File History to keep a good local backup of your system is highly recommended. For the cost of a single disk drive you will have a very reliable, automatic backup of your data.

Since file history is local to your system, it can be damaged or destroyed if your computer is infected by a virus.

System backup

Advanced Security Feature

The backup solutions above are great for data files such as documents, spreadsheets, videos, music, and so on. LiveDrive and Carbonite do a great job with these, and File History works well as a local backup of your own files.

However, none of these options protects your operating system or the applications themselves. If you spill water on your computer or the system hard drive (the C volume) crashes, you will need to rebuild the operating system from installation disks and reinstall all your applications.

You can (and should) keep all of the media (CDs, DVDs, etc.) in a safe place. Also print out and store the registration codes and licensing information from any software you install. If your system disk or computer is destroyed, you can then use the media to restore your software. LiveDrive or Carbonite can be used to restore your actual data after the operating system has been installed.

For more advanced users, Microsoft has a utility called System Backup which will create a copy of your system disk (and any other disks you'd like to include) for you. You'll need a hard drive, preferably an external drive, with at least double the space you need to back up. You can get 2tb USB3 drives for under a hundred dollars these days. That is a good size (and the USB3 means the drive will be fast) for just about anything.

Here is the procedure for backing up your system

1. Make sure your backup drive is installed and turned on.
2. Select the *File History* control panel.
3. Click the link at the bottom of the File History window called *System Image Backup*.
4. The first screen asks you which disk you want to use for the backup. Select the drive you installed for backups and click next.

5. On the next screen, click the drives you want to back up. I recommend you select the C drive and the one marked System Reserved (System). Then click next.
6. Look over the final screen. If all looks good click the *Start Backup* button.

This will take quite a bit of time (on the order of more than an hour) so it's a good idea to create these backups when you are not going to use your system for a while.

Some important facts about system backup:

- Restoring a system backup will completely overwrite the disk that you restore to. Thus if you restore your C drive from a system backup, the current contents of the C drive will be lost.
- A restore of your C drive can only be done when your system has been shut down. To restore you must boot up your system from some other device. The original installation disks are often used for this purpose.
- Restoring is an advanced procedure and if you have any doubts about how to do it get an expert to help you.
- Sometimes a system backup is the ONLY way to recover your system (other than a complete reinstall from the factory disks.) So even if you don't know how to restore it is a good idea to create one now and then. This gives you the option to do a restore if needed.

I recommend you create a system backup at least once a quarter and before you make any significant changes to your computer. When severe problems occur this may be the only way you can recover short of rebuilding the entire operating system and reinstalling all of your applications.

Best Practice

Create a system backup at least once a quarter and before any significant changes are made to your system.

System save and restore

Windows allows you to "take a snapshot", which is called a **restore point**, of the way your system looks at a particular point-in-time. This allows you to restore your operating system to exactly the way it was at the time the snapshot was created. This only affects the operating system (Windows) and not your own data files, documents, videos, and so on.

Windows will automatically create restore points now and then. Unfortunately, you cannot depend upon it doing so automatically, so it is best to create one before any major changes to your system.

Restore points are fully described on this web page.

http://goo.gl/SZJDCc

http://windows.microsoft.com/en-us/windows7/create-a-restore-point

The second Tuesday of every month is **Patch Tuesday**. This is the day when Microsoft releases modifications to the Windows operating system (called **patches**) to correct security issues as well as other problems they discover.

Best Practice

Create a restore point before installing anything on your system.

It is a good idea to create a restore point manually before you install patches. It is very rare for a problem to occur on Windows 7 or above due to patches, but it has been known to happen.

Using restore points is a fast and easy way to return your system back to a working condition if an install or patch update goes awry. If you install an application that makes your system act strangely, or one of the Microsoft Update patches fouls everything up, you can quickly get back to the way your system was before that time.

However, restore points will not protect you from some viruses, because a few of them actually modify or delete restore points. In the event of a virus which has done this kind of damage, you will need to rebuild your system either from the factory installation or from a system backup.

Follow the procedure below to create a restore point manually.

1. Select the *System* control panel.
2. Click on *Advanced System Settings*.
3. Click on the *Create* button
4. Follow the instructions.
5. It will take a few minutes to create the restore point.

To restore your system to the way it looked earlier, use the following procedure:

1. Select the *System* control panel.
2. Click on *Advanced System Settings*.
3. Click on the System Restore button.
4. Answer the questions that follow. You can select the recommended restore point or any restore point the system has on file.
5. The system will work for a few minutes, then reboot. Do not interrupt.
6. When the system boots up it will be restore to the previous state.

Check your backups occasionally

In the days just after Steve Jobs left Apple, our entire company switched from Macintosh computers to Windows PCs, except for one system. Sven was the man in charge of the plans for every single store in the chain. He claimed he didn't have the time to convert, didn't have any training, and, most importantly, the product he used to make store plans didn't work on a Windows computer.

Sven's Macintosh system was terribly old, yet it had been so reliable that no one even thought about it anymore.

He believed he was doing backups every day. Sometime in the past someone took the time to show Sven how to create a backup and he followed those instructions to the letter, even to the point of ignoring the error that occurred each time it ran. The instructions actually said, in writing, to ignore the error that occurred at that point in the procedure.

One day Sven's hard disk started making strange sounds, so he called the help desk. We tried to boot his system up but it was no go. We asked him if he was doing backups and he handed us his zip disks, which were blank! He had been faithfully doing backups for over two years, and not one of them had worked.

We had to send his hard disk out to a disk repair shop, and they managed to recover about 20% of the data at a cost of more than $6,000! It took the poor guy almost six months with two temps to get all the store plans recreated from scratch!

Be sure to check your backup, whatever product you choose to use, once in a while to ensure it is actually working. Not having a good backup when you need one can be a disaster of epic proportions.

Best Practice
Verify your backups are working properly when you first set them up and periodically after that.

Both LiveDrive and Carbonite have options to allow you to see what has been backed up to the cloud. You should check their reports occasionally, especially after the first backup has been completed. Look through the listing carefully to make sure it contains everything you expect. Remember, you may need to manually add music and video files to your backup. The documentation on each product explains how to do this.

Local backups that you perform manually should be carefully checked each time you copy the files.

File history has reports which enable you to see what has and has not been backed up. Take a look at them once in a while.

Private Internet Access

Advanced Security Feature

If you really want to ensure your privacy on the Internet, a product called *Private Internet Access* will help immensely. It's inexpensive, around $40 a year, and easy to install and use.

This application creates a tunnel (called a **Virtual Private Network**) between your system and one of the servers operated by Private Internet Access. This tunnel is secure, encrypted, and your IP address is hidden. By using this product you

---
Best Practice
Use a product such as Private Internet Access to provide a very high level of security on the internet.

---

are highly protected from people snooping on your data as it travels between your system and the Internet.

The web site for the product has all of the information about what this product does, how it works, and what protections it provides. I've been using it a while and have found it remarkably easy to use; simply download the product, install it, and click a couple of check boxes. The product does not appear to slow down Internet or system performance to any great degree.

To begin go to this web site.

http://goo.gl/OqCDnP

https://www.privateinternetaccess.com/pages/buy-vpn/

Scroll down to the various purchase options and choose the right one for you. Once you purchase the product you will receive two email messages in your email inbox. One contains the link from which to download the software. Just click on it and install.

The second contains the username and password for your connection. Make sure you've printed this out before you install.

Once the installation has completed a settings box will be displayed. Enter the username and password from the email you were sent. Change options if you want. It's a good idea to check "Start application at Logon" and "Auto-

Connect When App is Started" in order to make this fully automatic. You don't generally need to worry about the rest of these options.

Patching

Without a doubt, ensuring Windows Updates are applied each month is hands down the most important task you can perform to keep your system secure. Even if you do nothing else that is recommended in this book, just applying Windows Updates will improve your security immensely.

Let me repeat, the most important task you can perform to prevent any kind of malware infection is to regularly update your operating system *and* applications with updates (often called patches) from vendors. It doesn't matter which operating system you are using, Macintosh, Windows, Linux, or whatever, the most important thing to do to prevent infection is to regularly update.

> **Best Practice**
> *Use Windows Update to download and install software updates and set it to do so automatically.*

If a flaw is discovered and made public before it can be corrected by the vendor, it is called a zero-day threat. This means the vulnerability (or flaw) is known to hackers and can be exploited, but the vendor has not yet released the patch to correct the issue. Thus, patching is reactive in nature, meaning flaws are corrected after they are discovered. This is one of the reasons why you need to follow Best Practices and create a layered defense.

You see, operating systems are designed and written by people, and they occasionally make mistakes. Hackers write programs which attempt to take advantage of those mistakes to gain entrance to computers without permission. More importantly, hackers want to gain high privilege access (administrator rights are best) so their applications (viruses, Trojan horses, and so on) can do whatever the hacker wants to be done. There are hundreds or even thousands of these weaknesses scattered throughout an operating system, an application, or any other program running on your system.

The Windows operating system comes with an application called Windows Update which, by default, patches your operating system automatically whenever Microsoft releases patches, which is usually once a month. Many of these patches require a reboot to take effect, which is the reason why some people turn it off.

Regardless of whether you install patches manually or automatically, you must, if you want to be as safe as you can, ensure they are installed as soon as possible after the release date. Microsoft patches come out the second Tuesday of each month, meaning you should expect to update your system sometime during the second week of every single month. On rare occasions, Microsoft releases out-of-band patches – these should be applied immediately as they often address uber-critical vulnerabilities.

Why is it so important to not only to patch but to patch quickly? Hackers work fast, and as soon as they determine there is a vulnerability (a flaw in an operating system) they work on a way to exploit it. Sometimes, in what's known as a **zero-day exploit**, hackers are taking advantage of the vulnerability before the fix is released. The operating system vendor then has to scramble to get a correction out immediately. In this case, the fix is released, then usually re-released later because the first was done fast and ugly. The vendor then has to go back and ensure the patch works perfectly for everyone.

After you install Windows or when you first turn on your new computer, Windows Update is set to download and install patches automatically. You can check this using the Windows Update control panel. You have three options for installing updates.

- *Install updates automatically (recommended)* – This is the recommended selection. This will cause your computer to download updates and install them for you. Note that your system will automatically reboot once the updates are installed. However, you are presented with a popup message giving you one full day's notice before this reboot occurs.
- *Download updates but let me choose whether to install them* – If you want to manually perform the update yourself, you can save a lot of time by choosing this option. Your computer will automatically connect to Microsoft and download any updates to your computer. You can then install them at a convenient time. The advantage to this method is the downloading, which can be a very lengthy process, is done overnight while you are not using the computer.
- *Check for updates but let me choose whether to download and install them* – Choose this option if you want to control the entire process

yourself. You'll be informed that updates are available but the download and install will not occur until you manually make it happen.
- *Never check for updates (not recommended)* – This completely bypasses the most important task you need to perform for proper security. Don't choose this unless you totally understand and have a very good reason.

I highly recommend you set Windows Update to automatically download and install. This is the best and easiest option because all the work will be done for you.

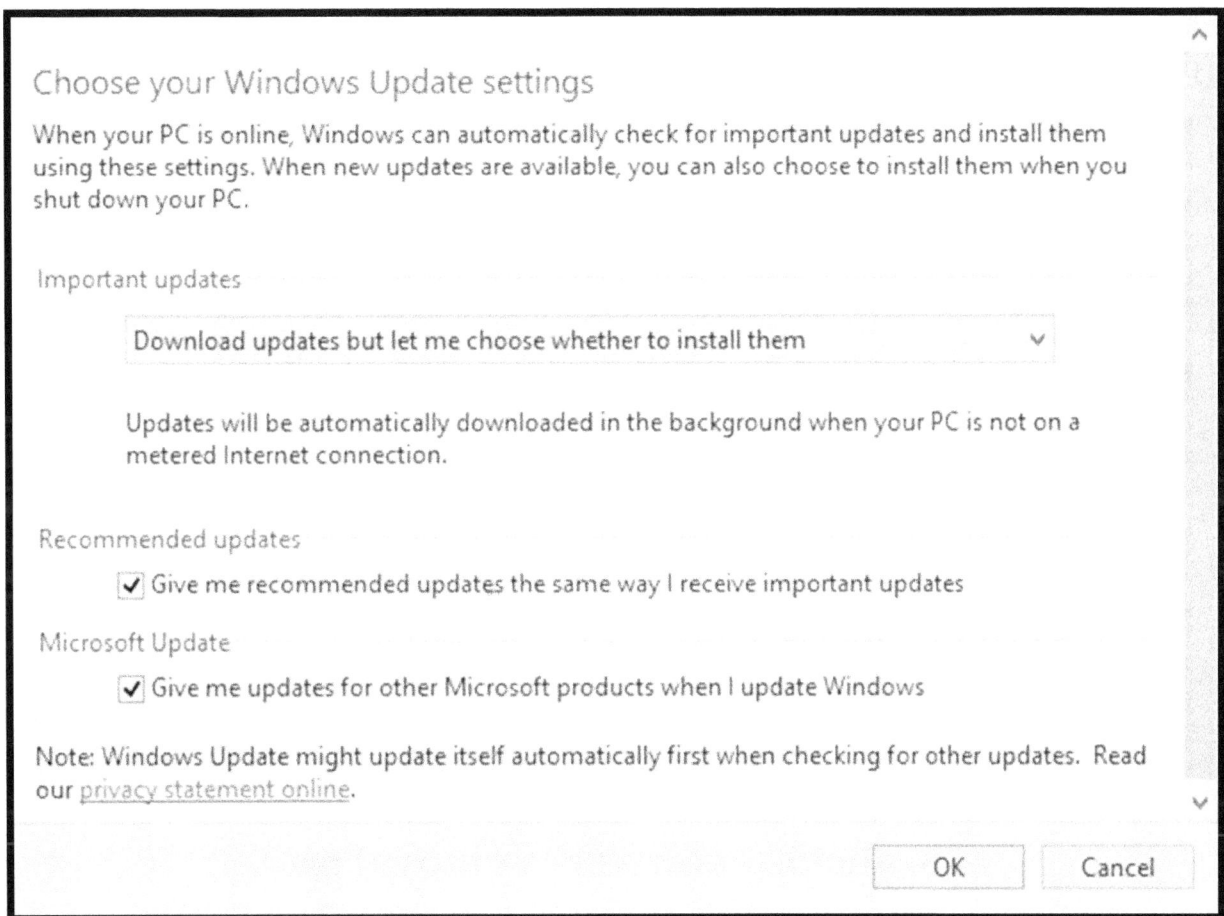

Choose your Windows Update settings

When your PC is online, Windows can automatically check for important updates and install them using these settings. When new updates are available, you can also choose to install them when you shut down your PC.

Important updates

> Download updates but let me choose whether to install them   ⌄

Updates will be automatically downloaded in the background when your PC is not on a metered Internet connection.

Recommended updates

☑ Give me recommended updates the same way I receive important updates

Microsoft Update

☑ Give me updates for other Microsoft products when I update Windows

Note: Windows Update might update itself automatically first when checking for other updates. Read our privacy statement online.

[OK]   [Cancel]

There are two other options on this screen.

- *Give me recommended updates the same way I receive important updates* – Sometimes Microsoft releases updates which are not critical but fix problems. Clicking this box will also install these as well as the important updates.

- *Give me updates for other Microsoft products when I update Windows* – Any other Microsoft products, such as Office, will also be updated at the same time as Windows.

Ensure both of these check boxes are selected to receive maximum protection.

If you choose to manually install updates, you'll be notified that updates are available. When you receive this notification go to the Windows Update control panel where you will see the number of updates that will be installed and a button that you can click to initiate the update process. Click that button and you'll see a window like the one shown below. This will display the update progress as it happens, then ask you if you want to reboot now or later.

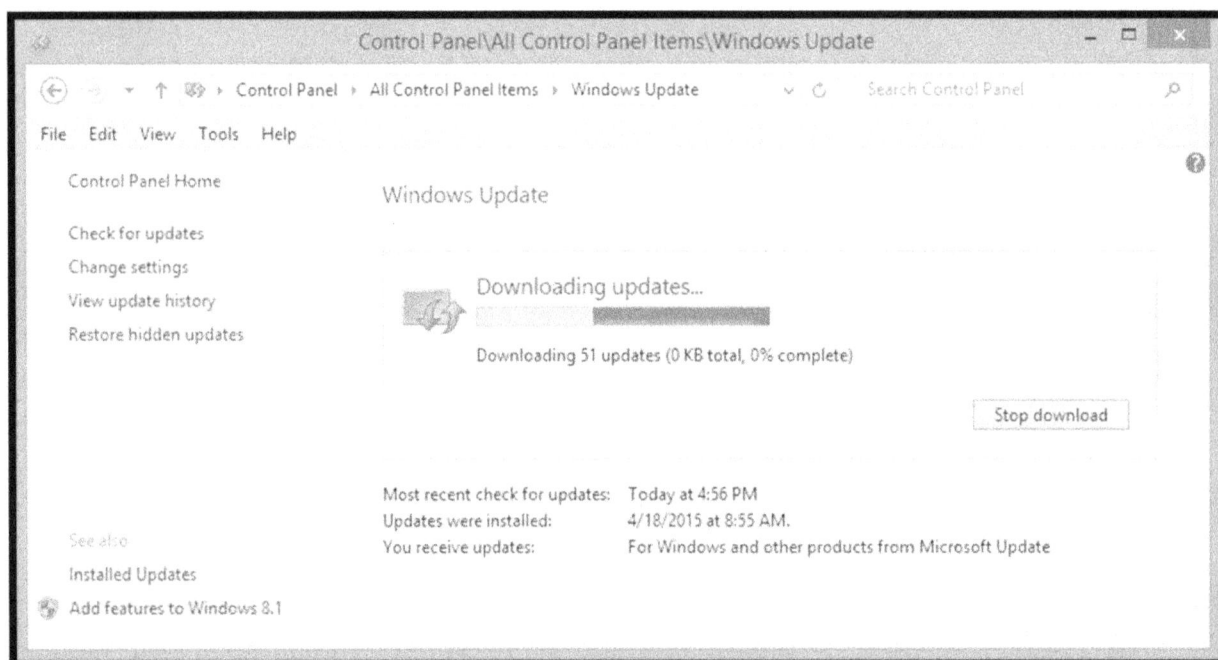

Malicious software removal tool

There are many known viruses that hide themselves in **very** hidden places all over your computer system. Naturally the person who wrote the malware wanted it to remain hidden for as long as possible to allow him or her to do as much harm as possible. The viruses will even recreate themselves if you don't manage to find and delete all of its bits and pieces.

That's where the Malicious Software Removal Tool comes into play. This handy little utility is automatically downloaded, installed, and run by Windows Update after each patch Tuesday. This is not a utility that you need to worry about as it is totally automatic.

Best Practice
Allow the Malicious Software Removal Tool to run normally during each monthly Windows Update cycle.

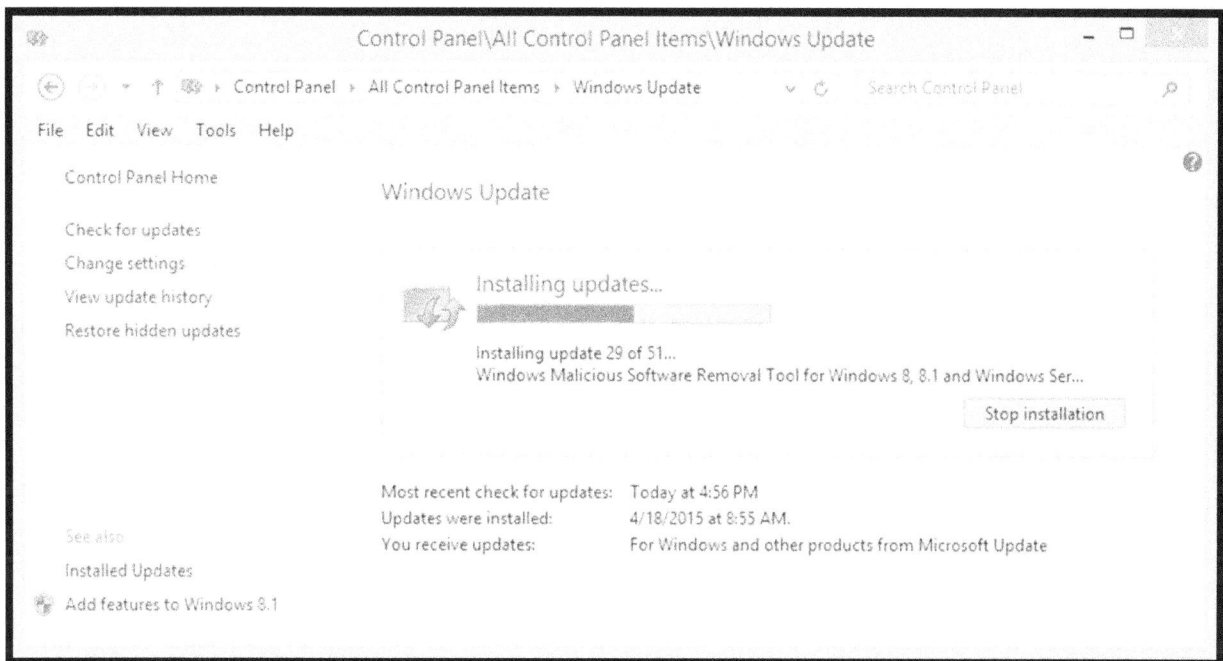

This application scans your computer looking for the traces of known computer infections. Those that it finds are removed completely and totally.

This tool automatically deletes itself when it has finished its task of scanning your computer. If you want to manually run it between patch cycles, you'll need to download it from Microsoft again at the following link. Note: part of the download process asks you to install the Bing toolbar. Be sure to uncheck the check box.

http://goo.gl/d7bQyK

https://www.microsoft.com/en-us/download/malicious-software-removal-tool-details.aspx

Once you've downloaded the scanner, simply click the image to run it on your system. You will be presented with a startup screen; just click the *next* button. The following window will appear

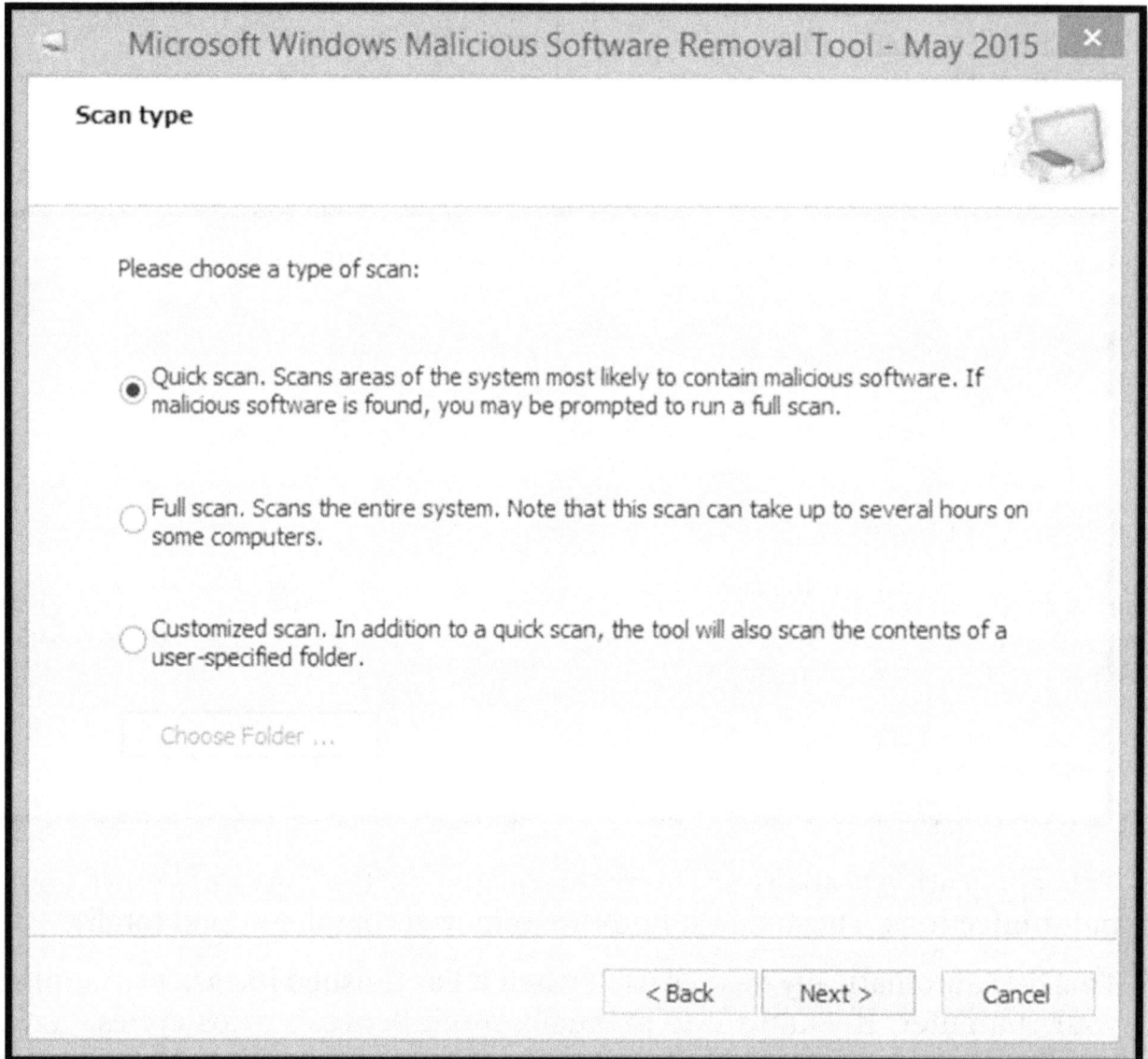

Microsoft Windows Malicious Software Removal Tool - May 2015

Scan type

Please choose a type of scan:

- Quick scan. Scans areas of the system most likely to contain malicious software. If malicious software is found, you may be prompted to run a full scan.

- Full scan. Scans the entire system. Note that this scan can take up to several hours on some computers.

- Customized scan. In addition to a quick scan, the tool will also scan the contents of a user-specified folder.

    Choose Folder ...

< Back   Next >   Cancel

You can select the quick scan (this is what is done during the monthly patch cycle) or a full system scan. Note that the full system scan can be quite lengthy.

Click the *Next* button to start the scan. A progress window will show you what the application is doing.

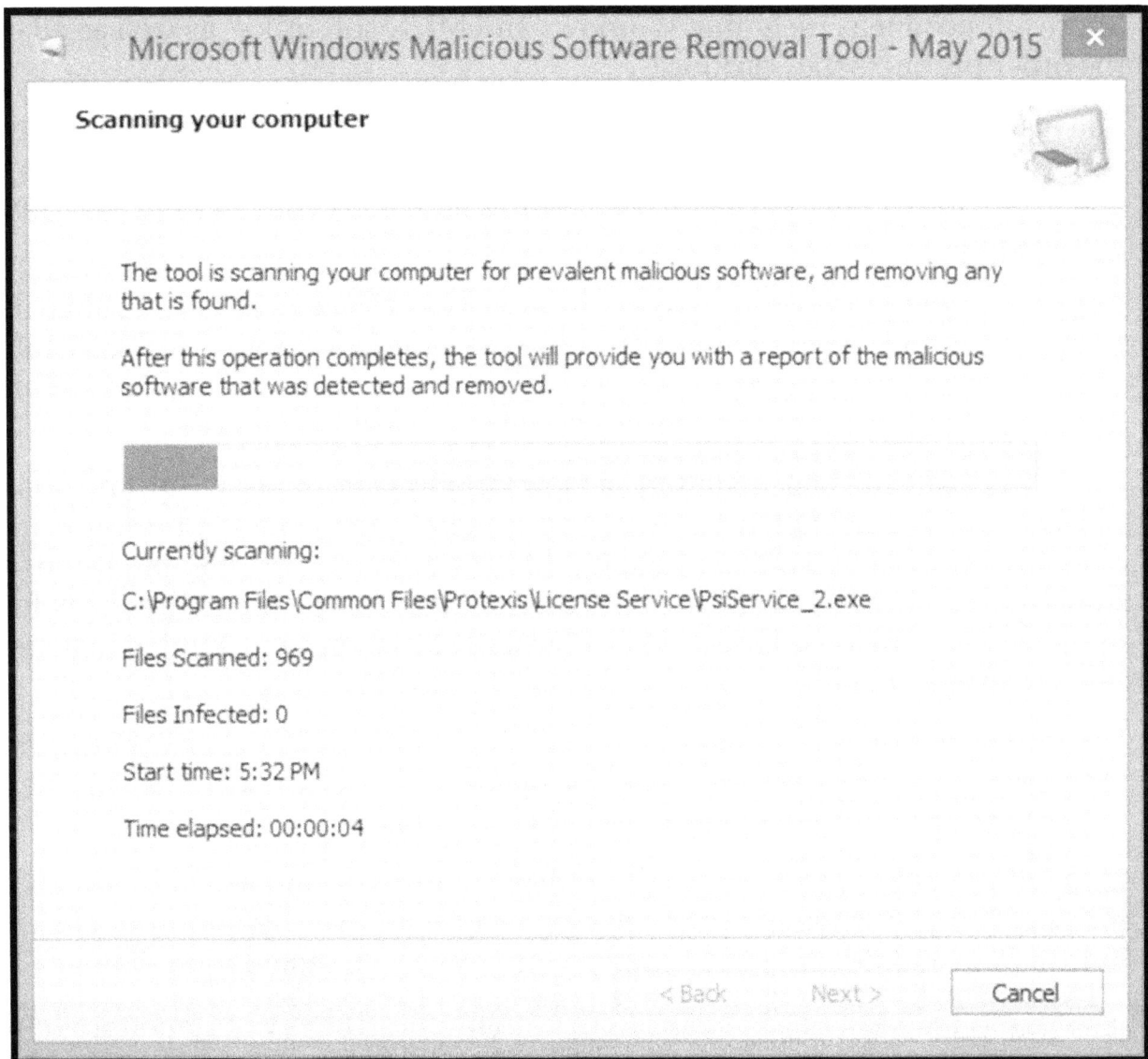

Why would you want to run this tool manually? If you suspect your system has been infected, running the tool manually can be useful to clean up any of the more common problems.

Updating applications

Microsoft Windows Update does an excellent job at keeping your system up-to-date to correct security issues. Unfortunately, it only handles updates for Microsoft products, including Windows and Microsoft office, as well as all of the other products they produce.

You also need to ensure security updates are installed for all the other applications that you have installed on your system. This must be done on a regular basis to prevent viruses from gaining access to your system via your applications.

Keeping applications up-to-date with security patches is an extraordinarily bothersome task. Some applications have options to help with this, while others only provide pages on web sites that you must check regularly. This differs from application to application and there is no standard.

Once you are finished installing a new application, start it up and look through the menus. Often you will find a menu item called "Check for updates" or something similar. Usually this is on the "Help" menu, although sometimes it is on the "Edit" or "File" menu. This lets you manually check for updates. If an update is found, you will usually be given some kind of instructions to install them.

You can also look under the "preferences" or "settings" in each application for a setting to automatically check for updates. Some applications will even have a setting to automatically install them for you.

Many of the better applications periodically check for updates and pop up a message for you when one is available. You then have the option to click a button to install, which usually walks you through an installation dialog.

To be completely safe, never click on a popup informing you that an update is required. Instead, open the application and do the update from there. Malware can attempt to install by causing a web page to popup an official looking message saying you need to do an update.

Unfortunately, many applications, especially those from smaller companies, do not have any options at all to check for updates. In these cases, you must visit their web site occasionally to see if one is available. Sometimes these companies will have an option somewhere on their web site (or when you install them) for you to subscribe to notifications of new versions. It is a good idea to select "YES" if you are asked if you'd like to be reminded of updates.

I like to download a trial copy of a product, if one is available, before purchasing an application. One of the things I look for in the trial is how simple it is to keep the product up-to-date. If there is no automatic update (or at least reminder) option, I typically will not purchase the product if there is any alternative.

Secunia

Fortunately, there is a solution to the problem of keeping your applications up-to-date. This is a nice little application called Secunia Personal Software Inspector (PSI).

To download this free product, visit this page:

http://goo.gl/hsohTZ

http://secunia.com/vulnerability_scanning/personal/

Once you've downloaded the application, install it on your computer. After the installation is complete, you'll see a small red or green icon in your system tray. Every day this will scan your computer to see if any applications require updates. The system tray icon will be red if updates are needed and green if none are required.

When the icon shows red double-click on it to bring up the Secunia PSI display. Don't be alarmed when it takes a few minutes to display on your screen. If all is clear (all applications are up-to-date) you'll see the following:

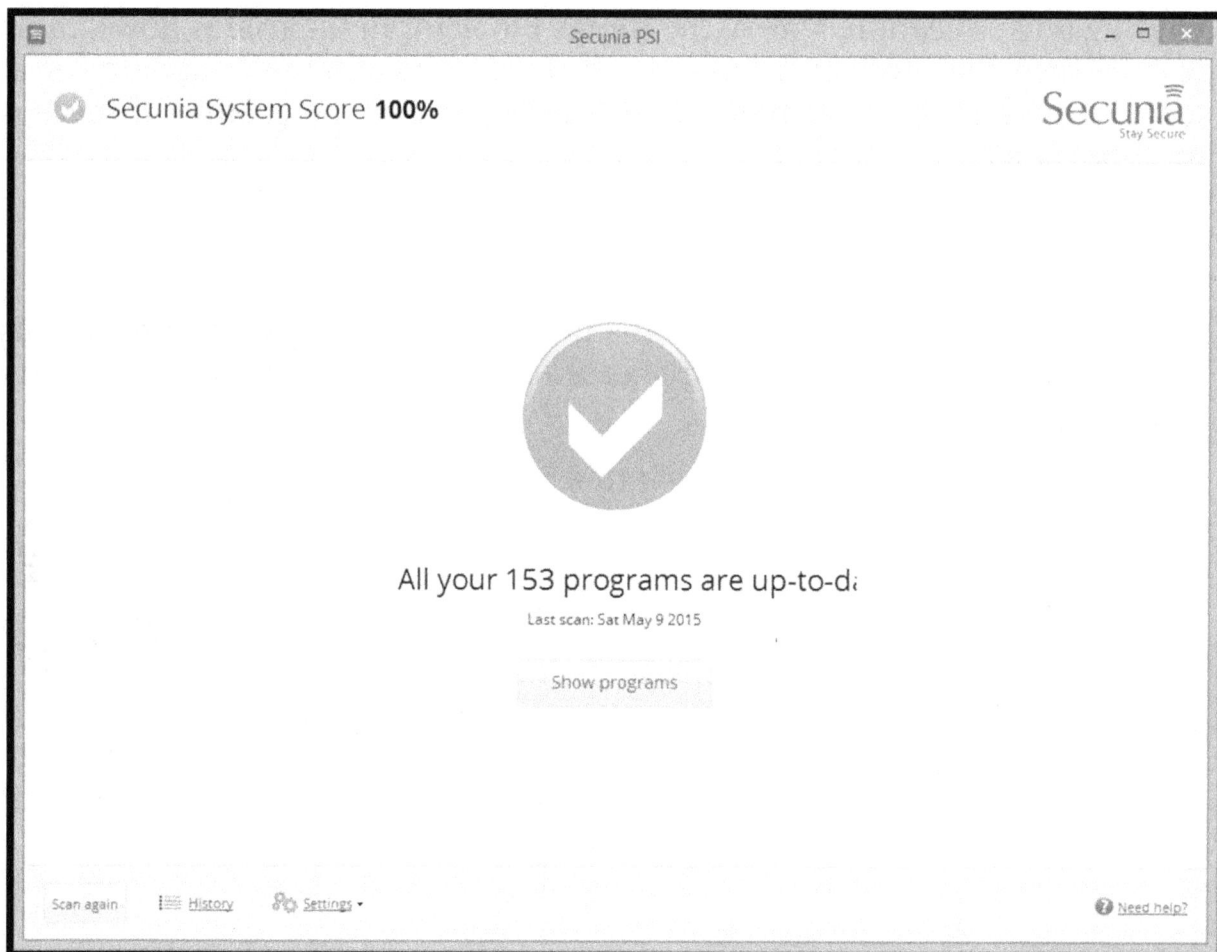

If any applications require updates you will be presented with a list of the programs that need updating. An option will be shown after each icon to let you know what you can do to correct the issue. You can press the "Scan again"

button at the bottom to cause Secunia to take another look through your applications.

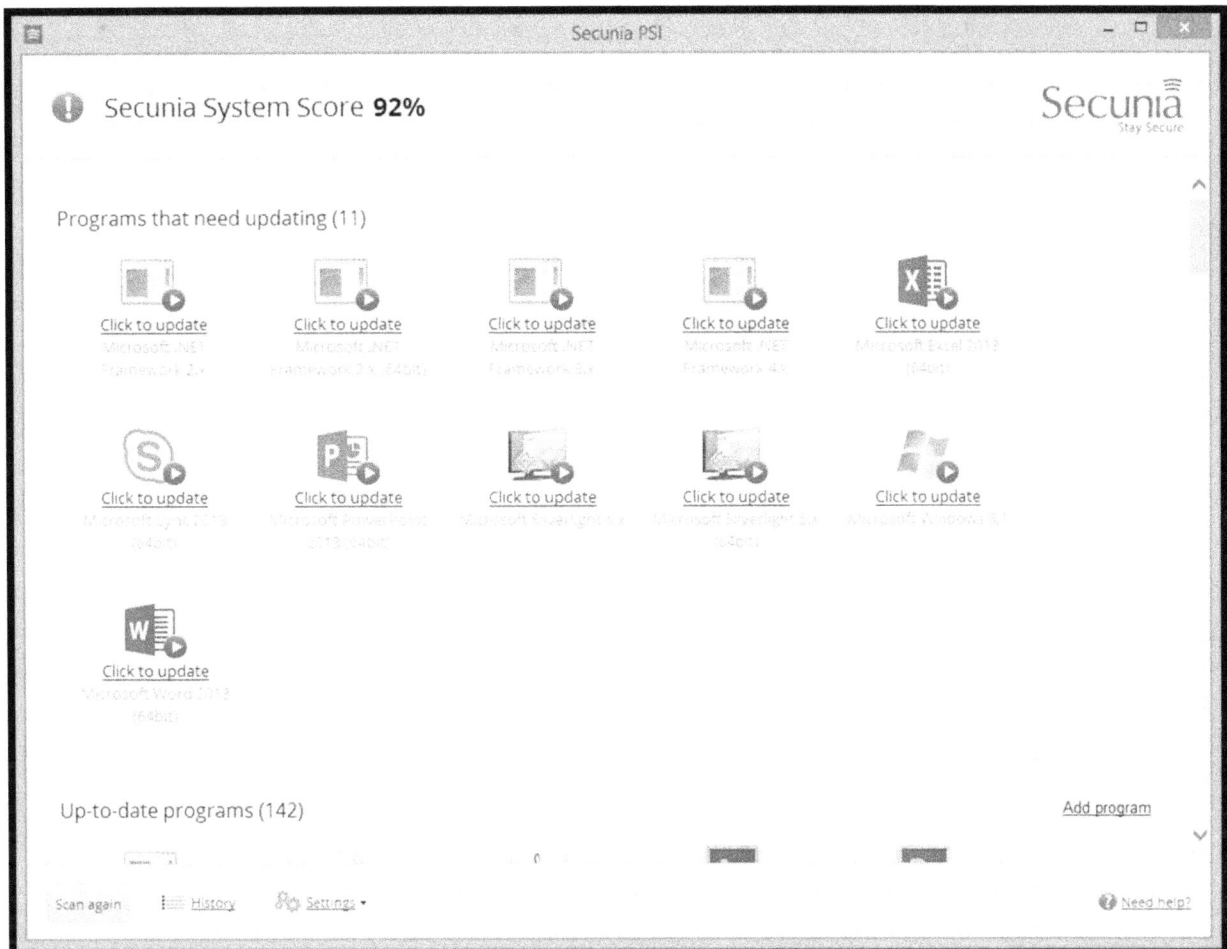

I've found the following procedure is best when using this application:

1. Run Windows update and install all of the updates.
2. Reboot (Windows update virtually always requires a reboot.)
3. Run Secunia.
4. Press the "Scan again" button.
5. Install any updates Secunia mentions.

Secunia does an excellent good job at finding applications that need to be updated. Unfortunately, sometimes it is unable to perform the update itself and you have to do it manually from the application.

The main value in using Securina is that it gives you a list of what needs to be updated so you don't have to manually check each application yourself.

Antivirus software

Some people claim, often with a sneer in their voice, that antivirus software is not needed. You'll occasionally hear this litany from users of Apple computers.

Others will claim that because they don't visit "bad neighborhoods" or pornographic web sites they are safe and don't need antivirus protection. I've even heard a few people say they don't want to run an antivirus application because it slows down their computer.

"Using a computer without running a good antivirus application is like not practicing safe sex. You might believe you don't need protection, but odds are you are going to get infected with something very nasty."

– Richard Lowe

Let me stress this lest there be any doubt. ***On ANY operating system, on any computer, it is essential that you install and run a good antivirus application***. All operating systems, including those produced by Apple, have vulnerabilities which allow malware to do nasty things to your computer.

Antivirus applications are based on the concept of signatures of known viruses. While this provides a very important level of protection it will not necessarily protect you from zero-day threats. This is one reason why a layered defense is so important. Malware can get through one level of defense but be stopped by another.

An exception to this rule is the Chromebook®. Good antivirus protection is built right into this operating system. Thus if you own a Chromebook this is one less application you will need to purchase and install.

In addition to vulnerabilities, sometimes software looks legitimate and enticing, yet it actually contains viruses. These can be downloaded from the Internet and installed. Because installations are normally done with administrator privileges, viruses and other malicious applications can be installed without relying on a vulnerability in the operating system.

If you are running a newer version of windows, Windows 7 or above, your computer hardware should have enough power to handle any antivirus application without issue.

Safe Computing is Like Safe Sex

I'm not going to talk about the different antivirus applications. There are quite a few of them and each has its own advantages and disadvantages. Over the years I've tried many of them, and for the most part they all do the job. When searching for an antivirus application choose one that has a Swiss Army Knife approach, meaning they include several security products in the same package.

A few of the antivirus applications that are available are listed below. Most of these products have free versions available. The price listed is for the professional version which includes firewalls and other security tools.

Product	Price	Link	Comments
Avast!	$40.00	http://goo.gl/4VHh8Y	
BitDefender	$90.00	http://goo.gl/SgEYTK	
MalWareBytes	$25.00	http://goo.gl/wjyzFG	Can be installed with other antivirus programs
McAfee	$60.00	http://goo.gl/m6KHWb	
Norton 360	$50.00	http://goo.gl/PFlR1f	
Panda	$50.00	http://goo.gl/NRMhAl	
Trend	$45.00	http://goo.gl/vbVzVP	

The following two antivirus applications have consistently worked well, and the good news is you can use them both at the same time.

Symantec Norton Security

Just about any of the anti-virus solutions listed above, and a number of others as well, will do the job required. I've been using Norton Security for several years and I've found it to be a reliable product. It contains a full security suite including a firewall, antivirus, antispam, anti-spyware, email scanning and browser protection.

You can install Norton by browsing to the following web page and purchasing the product. Follow the instructions to download and install.

http://goo.gl/PFlR1f

http://us.norton.com/norton-security-antivirus

Once you've installed the product, go into the extensions manager of Chrome and delete any of the Norton products you find. These Chrome extensions are less useful than other products recommended in this book.

Norton has a password vault. After extensive testing, I recommend LastPass instead. Norton's vault is adequate, but LastPass is a more specialized, better product.

The Windows applications listed below will be disabled by the installation of Norton.

- Windows Defender.
- Windows Security Essentials.
- Windows Firewall.

These applications are replaced by Norton.

By default Norton will protect your system from viruses, scan your web pages for threats, and perform many other functions to improve your security.

Malwarebytes

You can add another layer of protection to your system by installing Malwarebytes. This is an antivirus product that goes places that no other antivirus product will go.

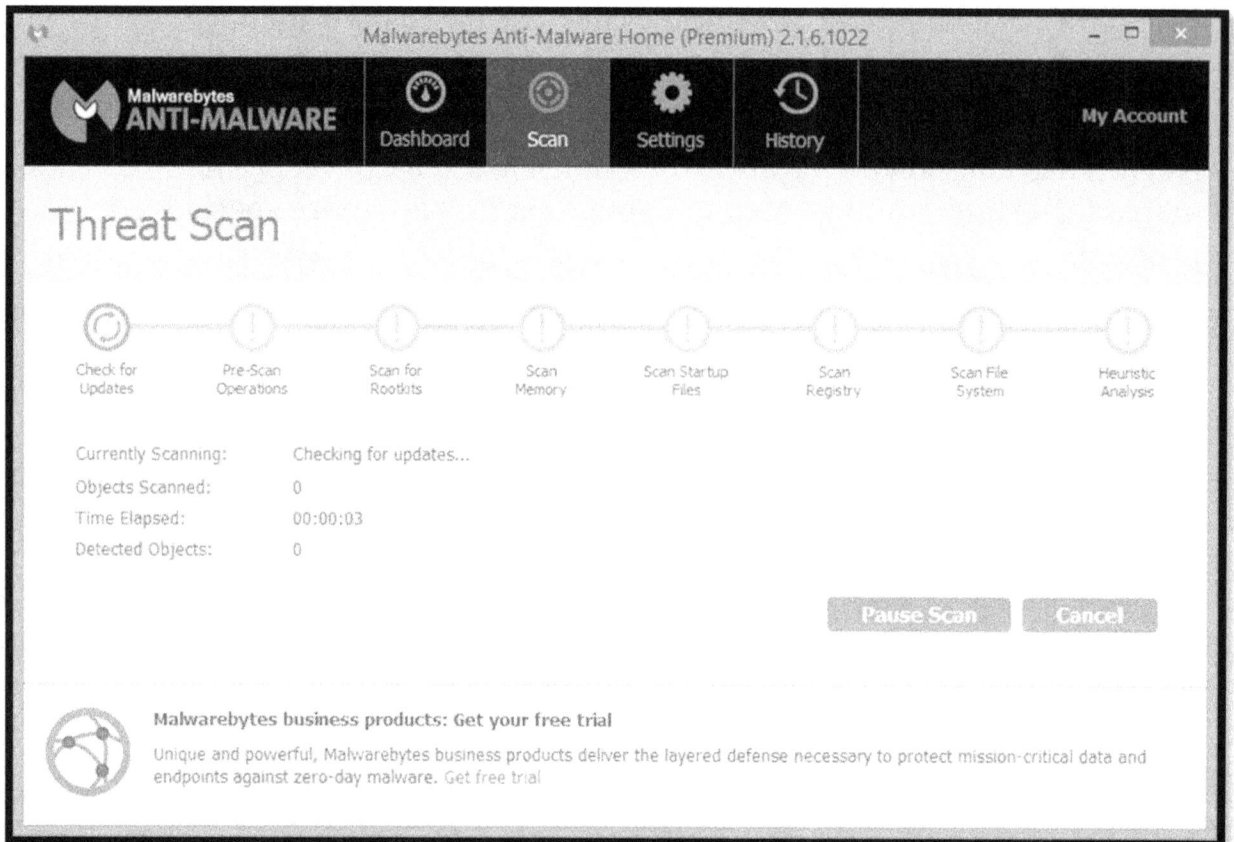

The product is available at the page listed below. You can install the free version or purchase the far more useful premium version.

http://goo.gl/wjyzFG

http://www.malwarebytes.org/products/

Installing Malwarebytes is simple. You just download it (and purchase the premium version of you'd like), install the application, and from that point onward it protects you from malicious software and web sites.

One of the primary reasons for spending the extra money for the premium version is Malwarebytes will then check each link that you click on to ensure it is not a known malicious web site. If Malwarebytes does believe the web site is malicious, a warning will be displayed. You will not be able to browse to that site unless you clock the "Exclude Web site" button.

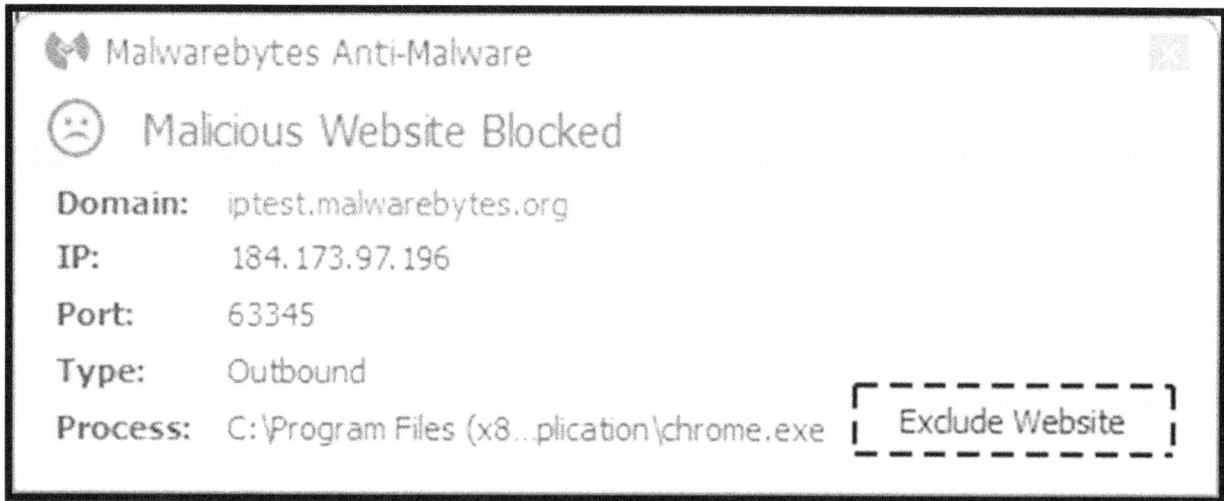

Malwarebytes is completely compatible with other antivirus products. You will occasionally, rarely, see Norton (or whatever other antivirus product you use) claim it cleaned up a virus in the Malwarebytes folders. Norton is simply seeing Malwarebytes data and mistaking it for actual threats. You can just ignore these messages.

Be sure the Enable self-protection module option is *not* checked (it is not by default.)

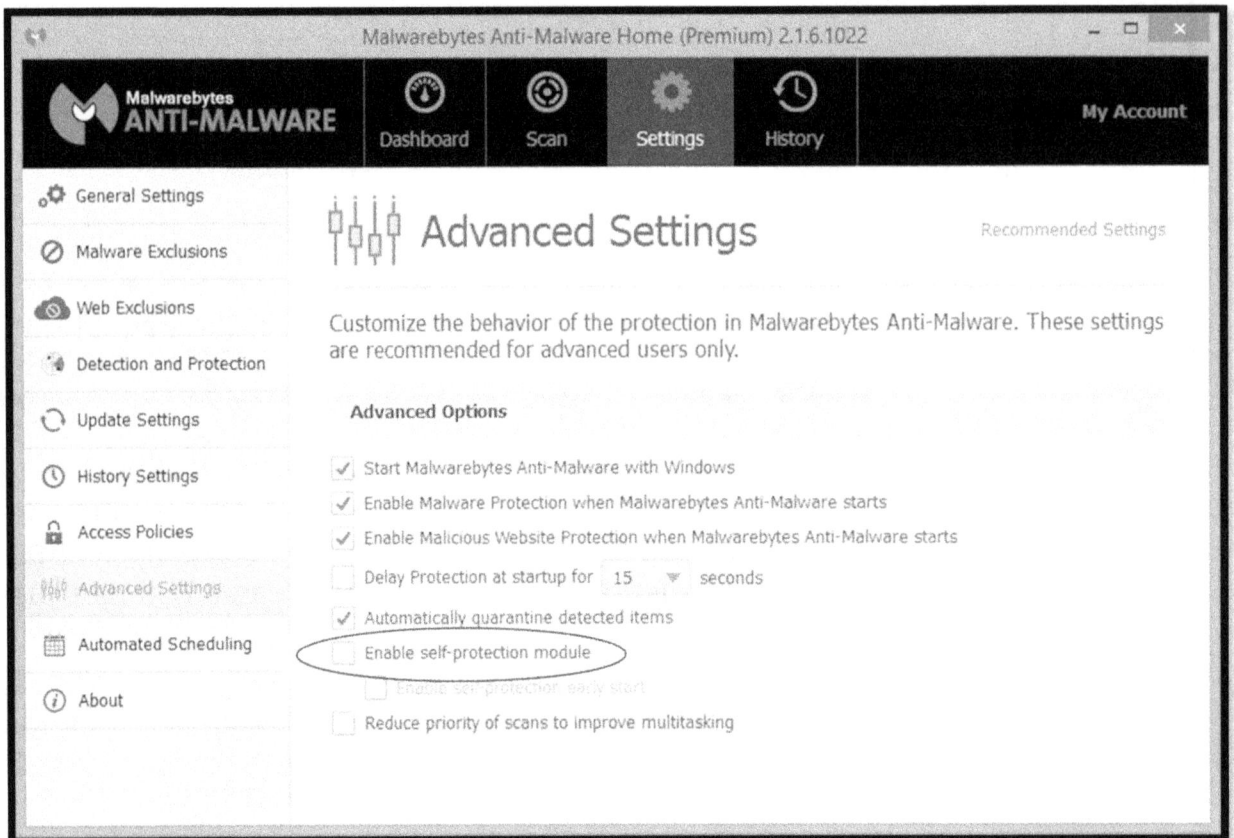

If Malwarebytes finds viruses on your computer it will repair any damage and put the viruses into a quarantine.

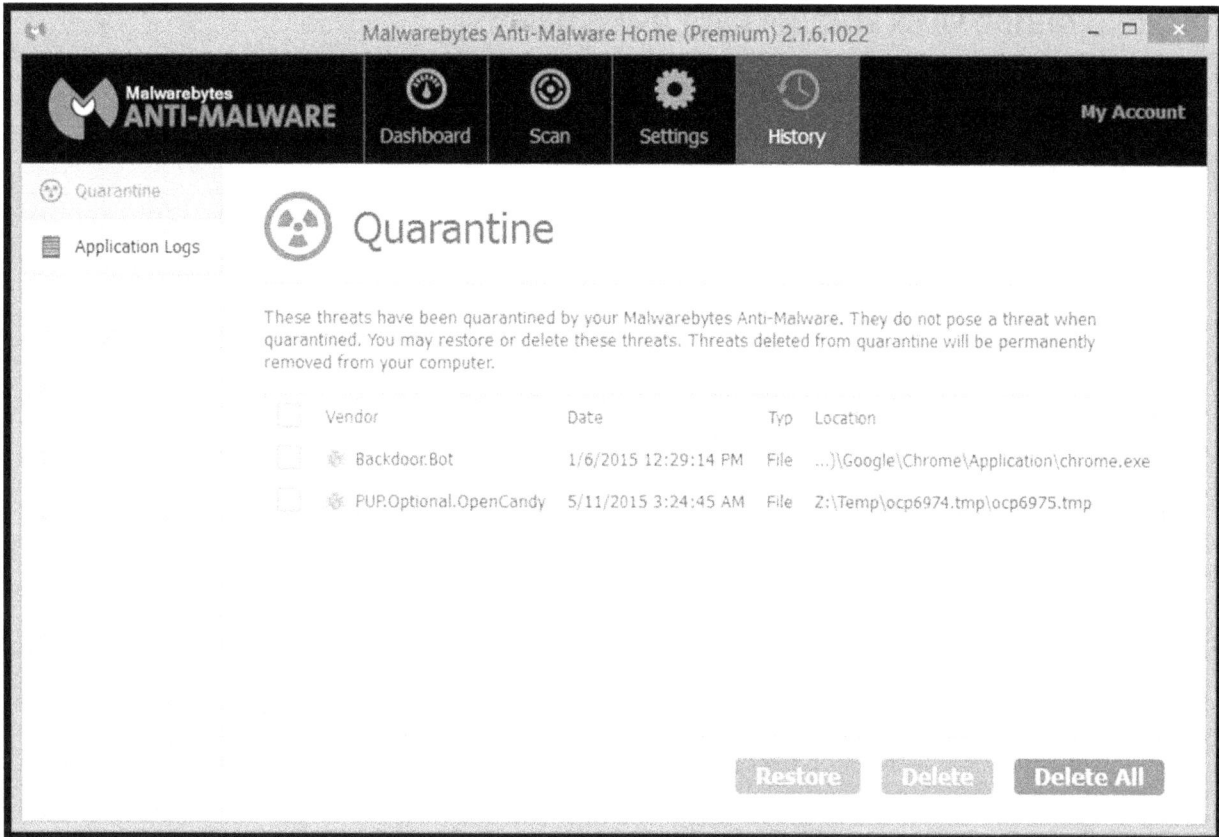

Online virus scanners

Many of the major antivirus companies provide online virus scanners. What this means is you can perform a virus scan on your system for no cost and without installing an entire antivirus product. Most of these options do require a download but they are much smaller versions than the normal packages.

These online scanners are very useful if you believe your system has been infected, and the infection disabled or corrupted the antivirus program you have installed.

You can generally run the online versions without concern for what is installed on your computer.

Several online virus scanners are listed below.

Trend Micro	http://housecall.trendmicro.com/
Microsoft	http://www.microsoft.com/security/scanner/
Kaspersky	http://www.kaspersky.com/security-scan
MacAfee	http://home.mcafee.com/downloads/free-virus-scan
Norton	https://security.symantec.com/nss/getnss.aspx

Firewalls

> *Many years ago I was installing Windows on a brand new computer. I hooked it up directly to the Internet so I could get to a site and download some updates. My system was only on the Internet for a few minutes, yet during that time it was infected by over a dozen serious viruses. I had to completely reformat the hard drive and reinstall Windows again.*

A firewall is a barrier between your network and the Internet that protects your system and network from mischief. The word firewall comes from an actual, physical wall built into a building to protect against fire. In the world of computers, these serve as an electronic barrier, always on guard and constantly resisting attempts to do your system harm.

You can think of a firewall as a large pane of glass covering the door to your home. It keeps insects, animals, and thieves from coming in without permission. Your basic firewall keeps things out by simply blocking the way in.

Windows comes with a firewall called, appropriately enough, *Microsoft Windows Firewall*. This provides basic protection against attacks from the Internet. Windows Firewall is generally good enough to protect the average home computer. However, if you want to do anything out of the ordinary you'll find it very difficult to use.

If you purchase a security suite such as Norton Security, the Windows Firewall will be replaced by the Norton Firewall product. This firewall provides much the same protection as the Windows Firewall, but is better integrated with the Norton Security product.

There are many firewall products that you can install on your computer to provide protection. If you do not purchase a security suite such as Norton Security, the Windows Firewall will give you adequate protection from intruders.

Locking down your system

Advanced Security Feature

There are two additional levels of protection that you can install or set on your Windows computer to make it even more secure.

- DEP, which stands for **Data Execution Prevention**. This is turned on for the Windows operating system and Microsoft applications. You can change it so it also protects from attacks within other applications. This is recommended. Note that occasionally (especially with games) this may cause an application to stop working. In that case you can add an exception to bypass the protection for that application only.
- EMET, **Enhanced Mitigation Experience Toolkit**, protects your computer from even more types of attacks. This is a program which you need to download from Microsoft and install. It is highly recommended that you do so.

It is very simple to add these extra levels of protection and the procedure is described in the following two sections.

DEP protects against damage from viruses

Normally DEP is turned on only for Windows applications, but you can improve your security by turning it on for all programs on your system. To do this:

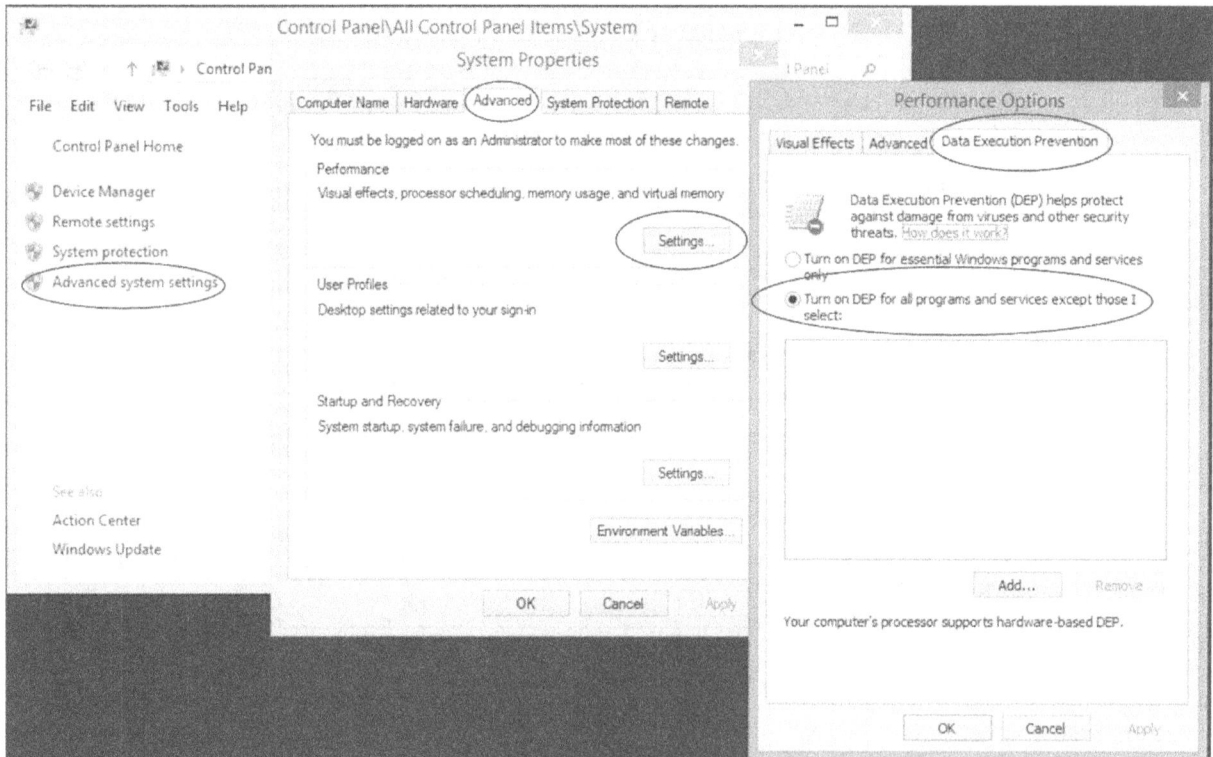

- Get to the System Control panel.
- Click on "advanced system settings."
- This will cause the "System Properties" box to be displayed.
- Click on the "Advanced" tab.
- Click on the "Settings…" button in the box labeled "Performance."
- This will display the "Performance Options" box.
- Click on the "Data Execution Prevention" tab.
- Now click the button marked "Turn on DEP for all programs and services except those I select:" Should this be a colon or a period?
- Click OK to save.
- Click OK in the System Properties box.

Now test your applications. Run anything on your system that you can think of including graphics editors, web browsers, email programs, video playing software, and everything else you've got. DEP does not normally break applications, but

> **Best Practice**
> Turn on DEP for all programs and services, then set exceptions for any that are affected by the change.

some older or poorly written programs can stop working or get strange errors. This is especially true of games, so check those extra carefully.

If you *do* find that a program breaks after setting this feature you can click the "Add..." button to tell Windows not to turn on DEP for that application.

EMET tightens down your system even more

EMET is a utility from Microsoft that you can install to make your system more secure. The concepts behind EMET are very complex, but all you need to do is download the program, install it and use the recommended settings. Once that is done your se-

> **Best Practice**
> Install EMET and select "use recommended settings" to improve your computers security.

curity will be vastly improved. Best of all, EMET is free.

To download EMET, visit this web page:

http://goo.gl/DfvdSE

https://www.microsoft.com/en-us/download/details.aspx?id=43714

Download the product, install it and click through the various windows to complete the installation. Once this has been completed, you will be presented with the screen below. Leave "Use Recommended Settings" selected and click finished. That's all there is to it.

Choose the download you want

File Name	Size
✔ EMET 5.1 Setup.msi	11.0 MB
EMET 5.1 User Guide.pdf	1.2 MB

Note the example below shows EMET 5.1 Setup.msi checked; the 5.1 part of this file name may be different for you.

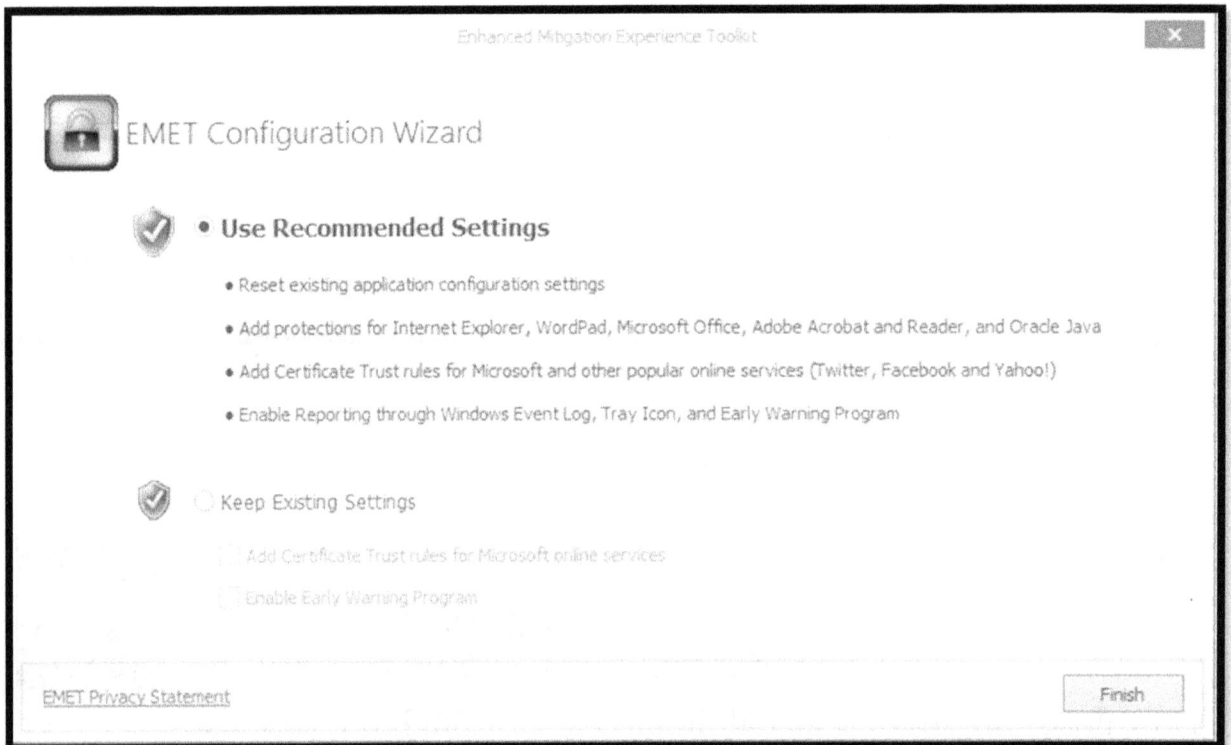

Enhanced Mitigation Experience Toolkit ☒

🔒 EMET Configuration Wizard

🛡 • **Use Recommended Settings**

• Reset existing application configuration settings

• Add protections for Internet Explorer, WordPad, Microsoft Office, Adobe Acrobat and Reader, and Oracle Java

• Add Certificate Trust rules for Microsoft and other popular online services (Twitter, Facebook and Yahoo!)

• Enable Reporting through Windows Event Log, Tray Icon, and Early Warning Program

🛡 ⦿ Keep Existing Settings

Add Certificate Trust rules for Microsoft online services

Enable Early Warning Program

EMET Privacy Statement | Finish |

EMET will stop any application that violates its rules. Because many viruses and other malicious applications will do things that violate EMET rulesets, they will often be stopped by EMET before they can do harm.

Google Chrome

The web browser you use impacts your security posture. Some browsers are more secure than others. Internet Explorer tends to have more weaknesses than other browsers. While you cannot uninstall Internet Explorer from a Windows system, you should stop using it if you can. Unfortunately Windows itself and its applications use Internet Explorer for various purposes.

> **Best Practice**
> *Do not use Internet Explorer to browse the web.*

I've found Google Chrome is a good option for web browsing. Download it from the following site:

http://goo.gl/c1yqAx

https://www.google.com/chrome/browser/desktop/

The Google Chrome web browser is absolutely free and is exceptionally secure. It automatically installs security updates on a regular basis, and has been designed and written with security in mind.

Google Chrome users

Once you install Chrome, one of the very first tasks you should perform is to add yourself as a user (or person as it is titled in Chrome). You don't have to do this, but it can help isolate the settings for each user. If you have two Chrome users, they can have separate settings, their cookies and other information is kept distinct and their browsing history is not mixed together.

For example, I had a visitor come by and I wanted to let her use my computer, but didn't want to create separate Windows accounts. This was because all she was doing was using Chrome to browse the web. Once I created the Chrome user she could browse anywhere on the web but couldn't see my history, bookmarks and other settings (nor could I see hers.) When she left I deleted her Chrome user and all her settings disappeared.

You should note, however, that Chrome users are not password protected in any way. Thus my friend could have easily seen all of my settings, browser history, and tabs, and I could see hers. In this case, that wasn't important. I just wanted to keep her settings separate while she was here.

You will find the Users option on first page of Chrome settings. To add a user, just click on "Add person"

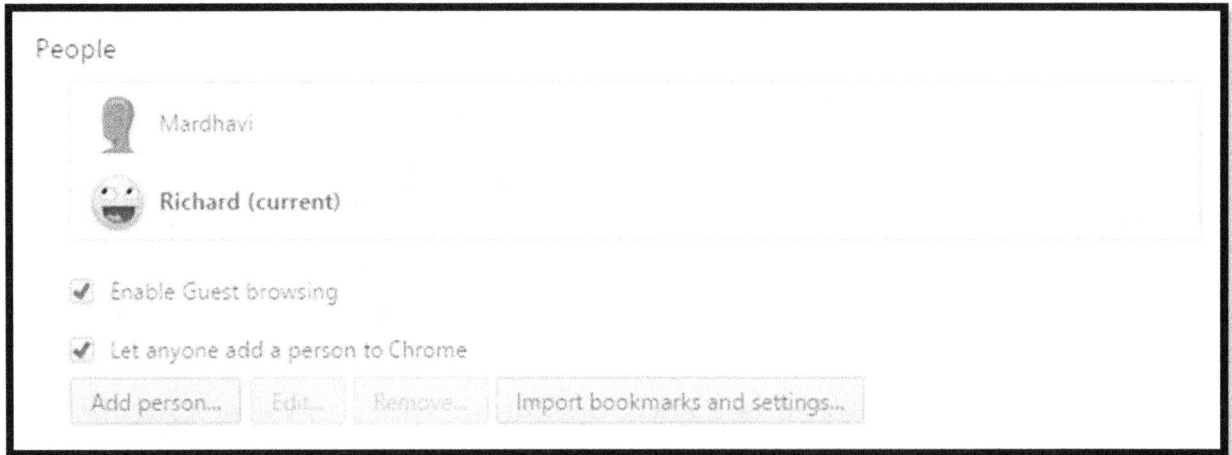

Clicking that button will display the "Add person" screen. Enter the information, click "Add" and you are done.

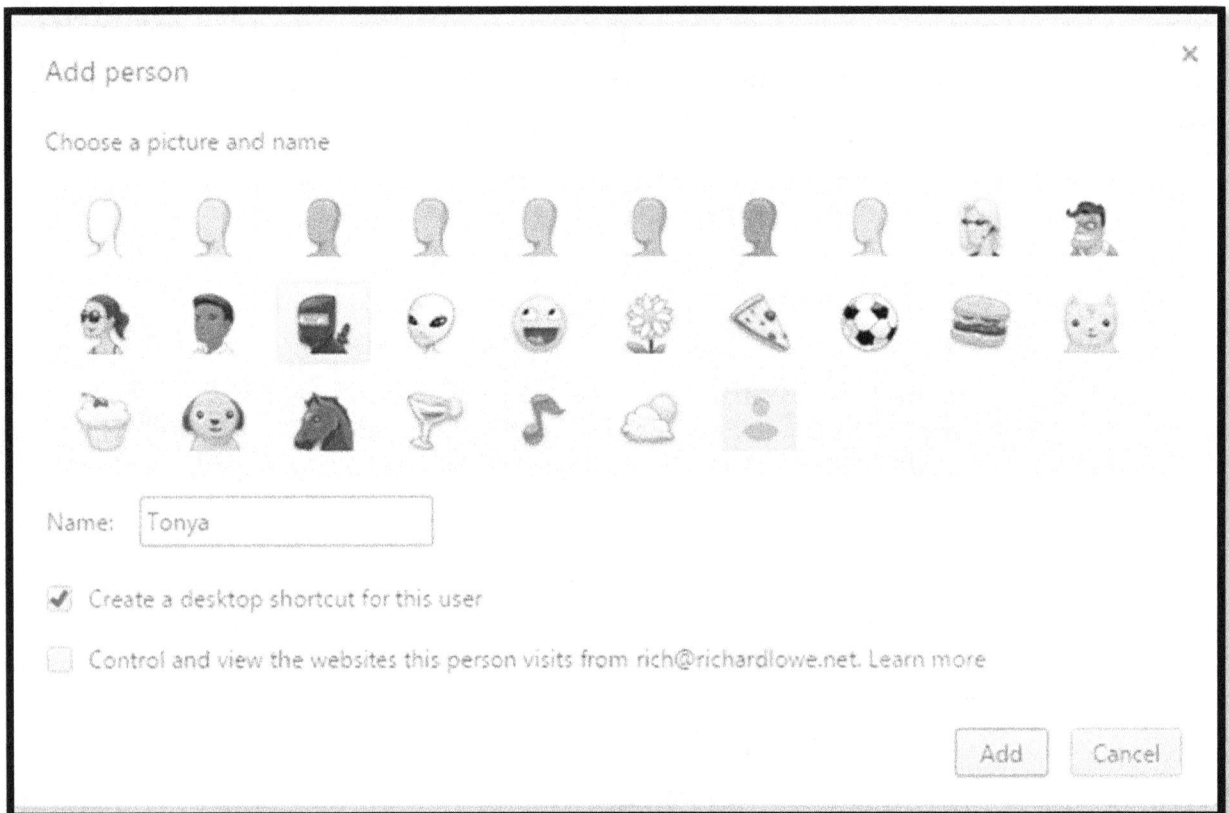

After you click "add," you will be prompted for the person's Google username and account. Signing in will import (if that person has it set that way in their settings) all of the settings information saved in their Google account.

Browser Extensions

There are thousands of browser extensions available for Chrome. Most of them are worthless, buggy, and some are even security risks. You need to be very careful which extensions you install. Don't go crazy. They often sound wonderful but by installing too many you will slow down your browser and possibly introduce security problems.

> **Best Practice**
> *Be very careful when installing browser extensions. Keep them to a minimum and only install those whom you trust either from a larger company or by references from others.*

Remember, many browser extensions can see *all* of your web activity, and some can even change what you see on web pages. So when you see that cool browser extension, ask yourself if you are comfortable with that.

Some browser extensions are extraordinarily useful for enhancing your browser security. I've researched quite a few of them, and these are the ones I have found to provide the best options for creating a safer environment.

- **Adblock Plus** – Remove advertisements from web pages.
- **Don't Track Me Google** – Turns off Google tracking.
- **Ghostery** – Turns off tracking and you can selectively turn it back on if needed.
- **LastPass** – Password manager.
- **LinkPeelr** – Expand short URLs so you know where they go.
- **Personal Blocklist (by Google)** – Prevent selected domains from appearing in Google search.
- **Privacy Manager** – Make the Chrome privacy controls more accessible.

Each of these is described in the following pages. If you install these and you find the browser slows down, you can uninstall them one by one until you have the performance you desire. The most important extensions are Adblock Plus and LastPass.

Adblock Plus

Best Practice

Install the Adblock Plus extension and use the appropriate filter lists to eliminate ads from your web pages.

Adblock Plus is one of the most insanely useful browser extensions ever created. It blocks all advertisements from your Chrome or Firefox browser.

The things it blocks include:

- Textual ads.
- Banner ads.
- Ads within YouTube videos.
- Tracking from social media sites such as Facebook, Google+ and others.
- Malicious software which is transmitted via online advertisements.

To install Adblock Plus, visit the following web page and follow the instructions.

http://goo.gl/2jDwn9

https://adblockplus.org/en/chrome

Once you have Adblock plus installed, you need to set up the filter lists.

The web page below has a complete tutorial on how to configure Adblock Plus after it is installed.

http://goo.gl/JlsXNN

https://adblockplus.org/en/tutorials

The page below has a complete list of all of the filters that are available. After you install Adblock Plus, look through the list and click on the "Subscribe" link for each list you prefer.

http://goo.gl/lxkMNB
https://adblockplus.org/en/subscriptions

By installing Adblock Plus and adding these filter lists (which are updated automatically for you as new advertisements are found) your browsing experience will be faster, more fulfilling and safer.

Don't track me Google

Each link on the Google search results really link to a Google tracking page which then links to the actual web page shown in the search history. This allows Google to track your web activities. It also requires more time since you first visit Google's tracking page, then visit the page you want.

This handy little extension removes that link to Google's tracking page and replaces it with the actual link to the web page you want.

Visit the page below and install the extension. That's all you need to do.

http://goo.gl/9E1l7C

Ghostery

This is a neat little extension which allows you to block trackers, giving you the option to just shut them all off or selectively choose just a few. This is very useful as some sites simply will not work if you don't allow certain specific trackers.

Visit this page to install the extension.

http://goo.gl/93YL4u

Once the extension is installed a wizard will pop up allowing you to configure the options. To move around the wizard there are left arrow and right arrow buttons. Simply follow the instructions and choose what options you desire. On the page titled "Blocking" I recommend clicking the "Select all" link to block all trackers.

Once you're finished with the configuration, the extension will display a small image of a blue ghost in the upper right of the Chrome browser. Click on this image to display your options for Ghostery. The screen shows each tracker with a slider. Red means blocked, green means not blocked.

If the site is not working you can click the "Whitelist Site" button and refresh the page. This turns off Ghostery for that page only so you can see if it is the cause of the problem. If the problem disappears, you can turn on the trackers one by one until you find out which one causes the page not to work properly.

Occasionally you will need to allow some trackers because video and music players will not play if their tracking is turned off. At first it can be a little confusing when a web page doesn't work, but once you get the hang of turning trackers on and off it becomes second nature.

LastPass

LastPass is a product which securely stores your passwords and usernames in a vault in the cloud.

To install it visit this page.

http://goo.gl/PF4DcR

You can install the free version or purchase the premium version for a few dollars a year (this is a very inexpensive product.)

LinkPeelr

This handy browser extension lets you hover your cursor over any shortened URL to see the longer version.

To install, go to this page.

http://goo.gl/1BWcTr

Once the extension is installed in Chrome, you can hover your cursor over any shortened link on any web page and see the longer version.

The example above shows how this works. This can be useful to avoid malicious, gambling, pornographic, and other sites you would prefer not to visit.

Personal Blocklist (by Google)

This very simple extension allow you to remove entries from Google search results. It adds a button beneath each search entry which removes the entire web site from your searches.

If a site constantly shows up in your searching that you don't want to see any more or you believe is risky, just click the button. You can edit the list if you change your mind later.

To install this extension, visit the following web page

http://goo.gl/PgVLko

The image below shows an example of the button added after search results.

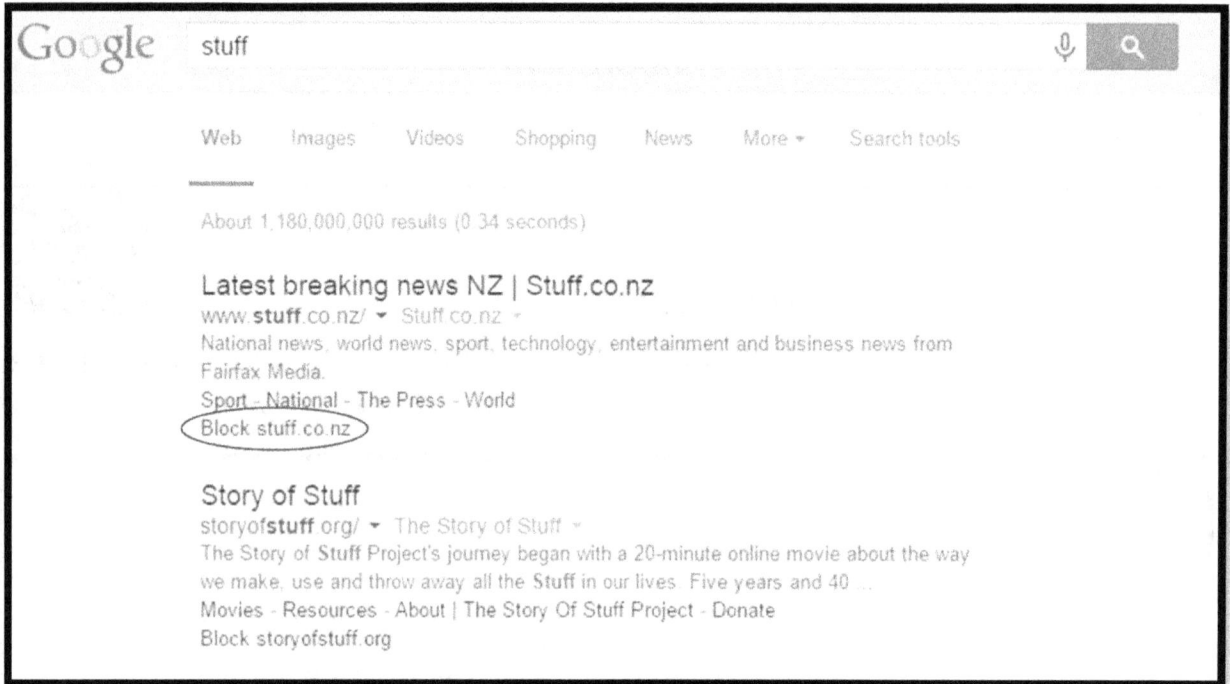

Privacy Manager

The privacy and security options for Chrome can be difficult to find and understand. This extension put all of the privacy options in one place, with handy on/off buttons so you can control exactly what you want.

To install this extension, visit the following web page:

http://goo.gl/fwyaoy

Once the extension has been installed, a small green icon will appear in the upper right corner of the browser. Click this icon to display all of your privacy settings, as shown below.

Each setting has a small question mark next to the name of the setting. Hover over that question mark to get an explanation of the setting. You can toggle each on or off.

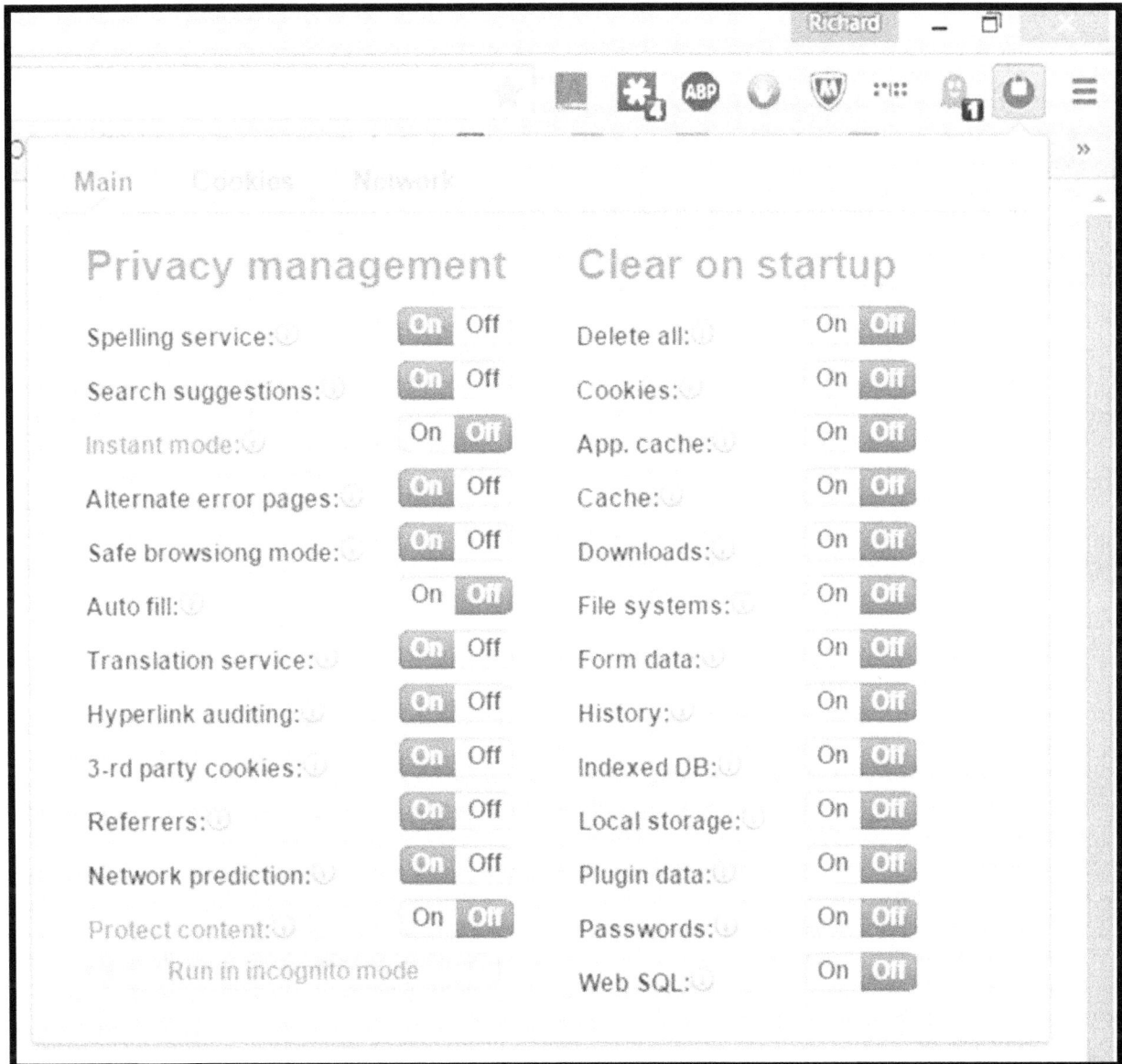

The "Clear on startup" column does exactly that for each item you have set to ON.

OpenDNS

Advanced Security Feature

Whenever you type in a web page name (a URL) your computer asks a domain name server to translate the name into an IP address. Generally your Internet provider sets you up with two of these servers, each with its own IP address, to do this translation for you. Your computer can be set up to automatically retrieve these IP Addresses from your Internet provider..

You can, however, use any DNS server that you want. This is literally true as DNS servers on the web, regardless of provider, may be used by anyone.

One of the very best DNS services for security purposes is known as OpenDNS. Using OpenDNS instead of the DNS servers from your Internet service provider has the following advantages:

Best Practice
Set up OpenDNS on all of your computers and your router.

- Several malicious web sites are blocked.
- Known phishing sites are blocked.
- You can define individual sites to be blocked.
- You can define classes of sites, such as gambling or pornography that will be blocked.

The web site for OpenDNS is shown below.

http://goo.gl/l1Cqfv

https://www.opendns.com/home-internet-security/parental-controls/opendns-home/

To enable OpenDNS, create an account on the OpenDNS web site. Choose one of the following options.

- OpenDNS Free – You can customize filters (which blocks web sites.) This also includes blocking of some malicious web sites.
- OpenDNS Home VIP – Same as free but costs $19.95 a year and includes support and reporting.

- OpenDNS Family Shield – Same as OpenDNS Free but is pre-config-ured to block adult web sites.

The free version is more than adequate for most people.

Next follow the instructions at the following web address.

http://goo.gl/EGquUb

https://store.opendns.com/setup/#/

You can enable OpenDNS directly on your computer or on your router. The above link has instructions for either.

The advantage of updating your router is *all* of the computers, smartphones, tablets and everything else on your network will use OpenDNS without any further settings. (Note this only applies while the device is connected to *your* network. If you connect at the local coffee shop, you are using their router and thus their DNS servers.)

The advantage of updating your computer directly is the OpenDNS servers will be used regardless of the network your device is connected to. That is to say if you bring your laptop to your coffee shop, you will still be using OpenDNS.

Cookies

With all of the rhetoric about cookies it is obvious that many people don't understand that these little text files were invented to solve a problem. In fact, cookies were created to resolve the Internet's equivalent of Alzheimer's disease. Web sites do not remember who they are talking to.

The web was designed to be simple and straightforward. You (through a browser such as Internet Explorer, Chrome or Firefox) ask for something (a video, image, text, web page, or some other object) from a web server. The web server obediently hands it to you, then goes off to do something else, completely forgetting about what is was doing before.

The web was never designed to support electronic commerce. It was designed to support reading text. Images, videos, sounds, and shopping carts were all shoehorned into the structure later.

Okay, so web servers are forgetful. What exactly does this mean? The browser asks the web server for an object and the server obligingly returns it. The connection to the browser is then closed and forgotten. (Note this is a simplification for illustration purposes.)

The next time the browser makes a request of the web server, it doesn't know that it is the same as before. As far as the server is concerned, every single request to do something is a unique inquiry from a different computer.

This makes any kind of transaction control very difficult. Think about it for a minute and you'll understand. You enter your personal information into a screen, which sends you to a second screen to enter your name and address. As far as the web server is concerned, these are two entirely different things. So how does the server understand that the address and the credit card data belong together even though they are two separate screens?

The answer is cookies. To put it very simply, a cookie is a computer's way of dropping bread crumbs - it's just a way for the web server to keep track of what it is doing. You could think of a cookie as a small notebook on your computer were web sites jot down things that they would otherwise forget. In the previous example, a cookie would allow the server to know that the name and address are related to the credit card number.

How does this work? Well, the server creates a small text file on your system called a cookie. This text file can only be referenced by that server, and it contains a simple unique number which identifies you.

Whenever you go to a web page, the web server first looks at the cookies that may have been previously recorded on your system. Thus, when a screen allowing you to enter your name and address is displayed, the browser tries to read a cookie, effectively asking "do I know who you are?" It does the same thing on the credit card entry screen.

Okay, this all seems harmless enough, doesn't it? So, how is this system abused?

Cookies can be set to last until the browser exits, or they can be set to expire (be deleted) far into the future. Various advertising companies actively abuse this feature – and this has led to the public backlash against cookies.

You see, cookies can be created and read when any object is loaded from a web server. This includes banners and **web bugs** (tiny images designed to help advertisers track who is looking at their ads.)

The advertising companies take advantage of this feature to set cookies on your computer so they can build up a picture of what sites you've been looking at. The banners effectively ask "have I seen this person (computer system) before?" If the answer is "yes" (a cookie exists), then a notation is made in your profile on the advertiser's computer system. It does not take long for an advertising agency to build up a very nice understanding of exactly what you do on the Internet. Why do they want to do this? To make more money, of course.

How does this work? An advertising agency sells eyeballs. The theory they operate on is simple. The more qualified the eyeballs, the more likely that banners will be clicked, and the more likely that sales are to be made. Thus, if you typically surf, say, Star Trek sites, you may be interested in seeing advertisements about Science Fiction movies, and theoretically you will be more likely to purchase tickets to those movies.

Okay, why is this a problem? Well, do you really want an advertising agency knowing everything about your web surfing habits? Do you trust them? Do you think they will keep this information private?

Or to put it another way, these companies are making money (and lots of it) based upon your eyeballs. They are not sharing that money with you; in fact, they never even asked your permission to gather this information.

As an analogy, suppose you were reading a magazine on a park bench and someone was hiding in the tree over your head, recording every page that you looked at in a notebook. How long would you put up with this behavior?

The public is simply objecting to the unethical use of cookies to track their movements through the Internet. And as you can see, a very useful tool has been corrupted in order to increase profits. This is the classic conflict of convenience versus security.

Built in Chrome cookie options

There are many options in Chrome within "Settings" to control cookies. When you first install Chrome it is set to allow all cookies, but you have the complete power to enable, disable, and delete them, and even to only allow cookies from specific web sites. This can get pretty complicated very quickly.

1. Select "Settings" from the Chrome menu.
2. Click the link "Show advanced settings…" at the bottom of the Settings page to see all of the additional options available.
3. Scroll down to the Privacy heading and click the "Content Settings…" button.
4. This will display the "Content settings" box.

5. The first options in this box are listed under "Cookies."

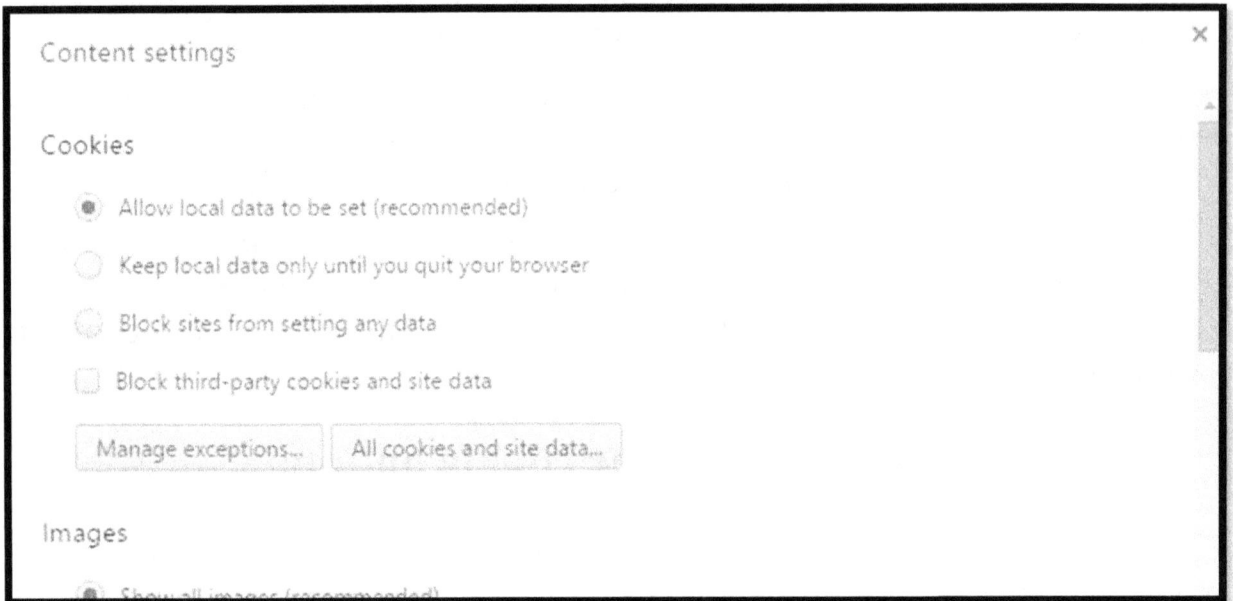

Content settings

Cookies

⦿ Allow local data to be set (recommended)

◯ Keep local data only until you quit your browser

◯ Block sites from setting any data

☐ Block third-party cookies and site data

[Manage exceptions...]   [All cookies and site data...]

Images

⦿ Show all images (recommended)

6. From here you can set the options as you require.

What's wrong with a little advertising?

We had a major virus outbreak, a major one. Dozens of computers were infected, and we didn't know where it was coming from. We had pretty strong security, so it seemed odd that we were getting hit all of a sudden. We sent samples of the virus to Symantec, and they were extremely helpful at identifying and cleansing the virus from each machine. However, as soon as we cleaned off one machine the virus popped up on another. This went on for about a month, and we were ready to rip our hair out in frustration.

It took some doing, but with some serious investigation we discovered the viruses were being introduced into our systems by web based advertisements. You know, those banner ads at the top and down the sides of just about every significant site on the Internet. Well, those ads can contain actual applets containing instructions to drop viruses on your computer system. This is an especially significant problem for computers which are not fully patched via Windows Update (or the equivalent on other computer operating systems).

B anner ads running on *any* site can be infected. The person who created the advertisement could have added the virus or the advertisement can get infected somewhere along the line. The point is any advertisement could potentially cause a virus infection.

There are two things you need to do to protect yourself from infected advertisements:

1. Ensure all patches for the operating system and applications are regularly updated.

2. Install an ad blocking application for the browsers that you use.

Each browser has its own method for stopping advertisements. My favorite is a little application called Adblock Plus. Once you install the extension advertisements will, for the most part, no longer be displayed. That particular method for receiving viruses will be stopped cold.

> **Best Practice**
> *Block web based advertising using AdBlock Plus.*

Browsing the web

If you are like most people you spend much of your day in a web browser. You might check out Facebook and your email first thing in the morning. After you get to work you may jump on the web to check the news, look at LinkedIn, and view some web sites about your profession.

It's become a fact of life - the web appears in every part of our lives. We play games on the web, perhaps using a desktop computer or an Xbox. We clip coupons, read the news, watch movies, upload pictures of our kids and vacations, write and receive email messages, tweet and Google, and we interact using Facebook.

Sometimes I wonder if people spend more time on the web than they spend actually talking and communicating with people in the real world.

Browsing the web is THE major way, in addition to email, that computer systems are infected with malicious software. Your web browser must be secure (Google Chrome is the best choice for this) and you need to use safe browsing habits. One of the best ways to remain safe while browsing the web is to be educated on the dangers and to always surf cautiously.

Securing your connections

Look at the URL in your browser. In the example below, this is the "www.microsoft.com" surrounded by the red circle. Notice the URL does not begin with https. This means your access is not secure, and a hacker, the government, or your company (if you are working on a computer at your office) can see the data you enter and the pages you view.

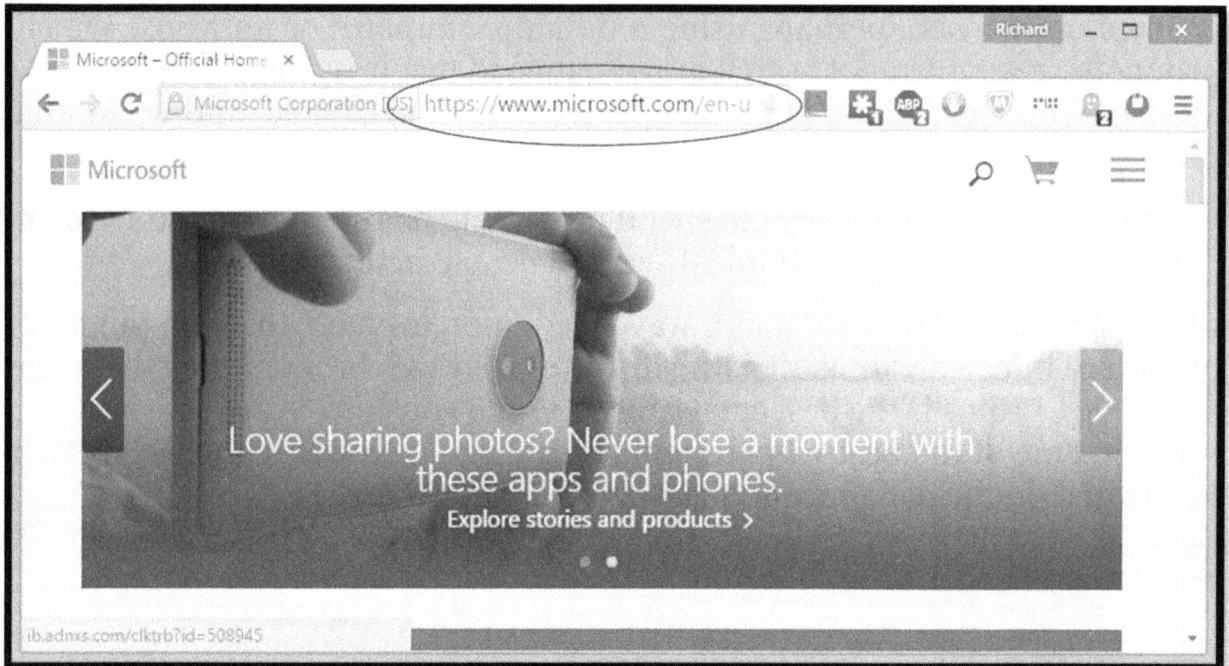

HTTPS means the web browser uses a very secure method to communicate with the web site, while the standard HTTP (which is often not displayed by the browser as you can see in the example above) does not use any security at all.

HTTPS causes the web browser to perform several important functions for you.

- It verifies that your web browser is actually communicating with the right web server. For example, if you browse to the Bank of America® web site, HTTPS will ensure you are actually communicating with the Bank of America web site. Note this does *not* protect you from mistyping the site URL.
- It ensures that the connection between your computer and the web site (web server) cannot be read by anyone else.

Note on the example below of the LinkedIn web site, *you do* see the HTTPS as part of the address (URL) of the web site. This means you can be sure, when using LinkedIn, that you are actually on the LinkedIn system and not a hacker's spoofed site, and that the information you enter on the web page cannot be intercepted by anyone else.

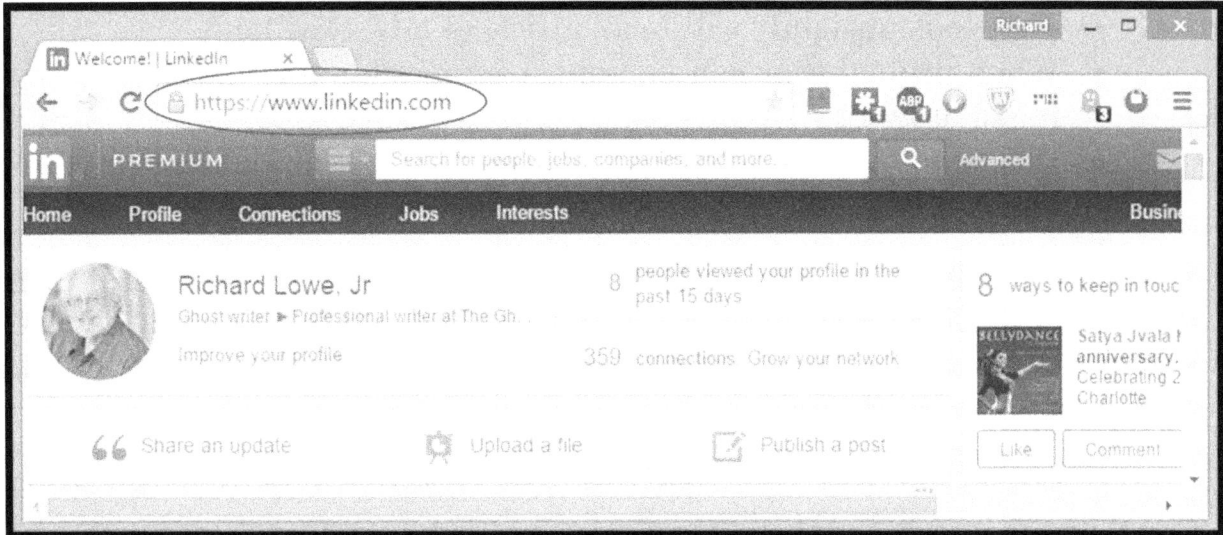

I want to stress how vital it is that you ensure the sites you use for personal and private data use HTTPS. When you use social media, such as Facebook, make sure you are using HTTPS (Facebook now does this automatically.) If you are using a shopping site such as Amazon.Com, ensure it is using HTTPS.

> **Best Practice**
> *Always use HTTPS when you are entering any personal information. If the web site doesn't support HTTPS stop using the web site.*

When you use Amazon, you will notice that as you browse through the pages it is not using HTTPS. This is fine. What you want to ensure is that HTTPS is being used when any sensitive or private data is being entered or received. In the case of Amazon, you will see HTTPS is used when you are looking at or changing account data, your shopping cart, and any other areas containing sensitive information.

Tiny URLs

URLs can get very long. In fact they can be so long as to become virtually unusable by normal human beings. Some of the problems caused by longer URLs are listed below.

- They look unsightly within an article or advertisement.
- When printed they can be almost impossible for someone to type into a browser.
- Web sites such as Twitter have a limited number of characters and longer URLs won't fit.

There are a huge number of URL shortening services available. All of them translate a short URL to the web site address (URL). One of the best services is provided by Google, at the following web site. This is the one used for URLs in this book.

http://goo.gl/

Google's site is very simple to use if you already have a Google account.

For example, you can use it to translate the following URL

http://www.richardlowe.com

To a shorter one:

http://goo.gl/XmyXKw

Useful as they can be, these short URLs do introduce a problem to web browsing. They hide the actual web site from you. The site could be a normal, everyday site, it could be a pornographic site, or a site trying to get you to download malicious software.

What should you do about this? You can install a browser extension to translate the shortened URL to its longer version. Linkpeelr is a browser extension which performs this function.

Web sites to avoid

> *Bob had an older computer running Windows XP. His system was not up-to-date on security updates. His system had an old modem (a device which can use a phone line to dial up the Internet or make phone calls directly.) Bob discovered that the Internet is full of free pornography and he went a little bit wild, visiting site after site after site. One of those websites downloaded a small malicious program, a very old virus, which dials 976 numbers from the modem attached to a computer. By the end of the month poor Bob had a phone bill for thousands of dollars from the malware literally making hundreds of calls to those pay-per-call phone numbers.*

Yes, I know some people love their pornographic web sites, while others tend towards online gambling. Some like to check out hacker sites and others look for codes to hack their games.

The problem with these types of web sites is they are often completely infested with malware of all types and varieties. They often include drive-by downloads of malicious software, meaning software that can be installed just by viewing a web page. In addition, they have lots of flashy applications which literally beg to be installed on your system. Each one of these applications is potentially dangerous to your computer.

Pornographic web sites

If you really ***do*** want to infect your system with a virus or other malicious program, just go ahead and visit lots of pornographic web sites. As you download pictures, videos, games, and other things from these sites, you're also most likely downloading malicious software. Simply viewing a web page may even install malware on your computer.

> **Best Practice**
> *Do not browse random pornographic web sites. If you do so your computer will be infected with malware.*

If you have some compulsion towards viewing pornography on the Internet you'd be best advised to get a cheap computer, something like a Chromebook, and use it only for your habit. Keep it off your more expensive Windows or Apple computer (yes, this effects both operating systems.) And be prepared to constantly reset that computer back to factory default settings as it becomes infested time after time.

If you habitually visit different pornographic web sites, especially free sites, your computer *will* be infected with malware.

Online gaming and gambling

"Online gaming sites are a major distribution vehicle for malware. Malware payloads target specific games.

Be very careful about using your computer to play online games and online gambling. They are often infected with malicious software which can download itself to your computer without your knowledge. This puts you at risk for phishing, identity theft, fraud and even straight theft. As with pornography, if you must use these types of sites get a cheap computer if you can.

The larger online gaming sites (such as Xbox) are generally pretty safe. Less well-known sites should be considered suspect. Avoid online gambling web sites entirely.

Click bait

The headline was very enticing. "3-Year-Old Boy Sneaks Into 90-Year-Old's Yard. You'll NEVER Believe What They Did Next! I'm In Tears!" When he clicked on the article it was just about a 3 year old moving away from his 89–year-old grandfather. You fell for Click Bait.

If you've been on Facebook or one of the other social media sites, then you are undoubtedly familiar with clickbait. These are headlines which are designed to get you to click. They are usually untrue and are just a way to entice you to view a web page or a video.

Why do they do this? Because those videos and web pages contain advertisements. Sometimes the sites make money just because you looked at the page. More often they cash in big when you click on the advertisements.

Clickbait is pure and simply a scam. The idea is to get you to move your eyeballs to a web site you probably wouldn't normally visit.

Here are a few examples of clickbait from a quick review of my Facebook page today.

- I get A+, Vegemon! Can you identify the vegetable?
- 20 Things to Remember If You Love a Person with ADD.
- If Your Friends Ever Say They Have ADHD, Just Show Them This.
- Groupon Posted This Product On Facebook, And The Comments That Followed Are A Masterpiece.
- A Farmer Was Drilling For Water When He Found Something That Shocked The Entire World. Seriously.

These headlines are all your basic con jobs. They push emotional buttons or pretend some vast importance or urgency that does not exist.

What do you do about clickbait? Simple. Don't click on the link. Don't share the links on social media sites such as Facebook. Don't comment, not even to point out that it is clickbait. Simply ignore the link.

Likejacking

The picture of a deformed baby had a headline which read "Facebook will donate a dollar for every Like!" Bobby was happy to help. After filling out three surveys though, he wasn't so sure. It seemed like an awful lot of work for a one dollar donation. Six month later when he examined his credit reports he realized those surveys had enabled some criminals to steal his identity and open up a dozen credit cards in his name.

Have you seen any of the following on Facebook or other social media sites?

- Offers of free giftcards.
- Bill Gates offering money.
- Offers for free iPads, iPhones and so on.
- Pictures of a sick child.

These are almost certainly attempts to **LikeJack**. They are attempts to get you to click a link. If you do, you will often be asked to fill out surveys and applications of one kind or another. There are several purposes for this:

- To directly earn money from filling out surveys and clicking on advertisements.
- To steal your personal information for identity theft purposes.
- To steal your credit card numbers.

- To install an application which spreads the post to all of your friends.

In addition, since you clicked the Like button, all of the people in your friends list will see the post and can also Like it. Because these posts have very catchy titles they tend to go viral very quickly.

Facebook and other social media sites have done a lot of work to cut down on this kind of thing, so it has become less of an issue that used to be. The criminals keep trying, however, and occasionally figure out how to break through the defenses.

So how do you avoid falling for LikeJack schemes? Just avoid them. If it seems too good to be true, it is almost certainly a lie.

Likebait

LikeBait is the same as ClickBait, only the idea is to click the Like button instead of clicking on a link. Personally, I tend to avoid the Like button entirely. If you enjoy Liking things, my advice would be to use it sparingly.

Warez

Lester thought he had it made! His friend told him about this great site which had free software to download. Of course Lester wanted to save money, so he installed the software on his system. A week later he realized that was one of the most expensive free things he'd ever purchased, as his computer got heavily infected. He had to hire an expert to recover his data and rebuild his system, which cost several hundred dollars.

Some hackers have discovered they can get around the license requirements of applications such as Microsoft Word and other products. They often make these available to others on the web. Sometimes these are posted for free download and sometimes they ask for a payment. These applications are called **warez.**

Best Practice
Do not install illegal copies of applications.

Never download warez. Never visit warez or hacker sites. First, downloading and installing warez is illegal. Second, warez applications often contain malicious software such as viruses, Trojan horses, and worse.

Merely visiting warez or other hacking and cracking web sites is likely to infect your computer with viruses. Downloading and installing warez applications almost certainly will cause infection, and is illegal as well.

Incognito

One of the easiest ways to maintain your privacy is to browse using *incognito*. This works exactly like a normal Chrome browser window, except that when you close the last incognito window all cookies, browser history, and search history that occurred while you were in incognito is deleted.

I often keep two browser windows open (this is easiest if your system has two monitors.) One of the windows is *not* incognito and the other one *is* incognito. Any sites that require cookies, such as my bank, are accessed from the non-incognito window. Any searches or web surfing are done from the incognito window.

It is important to understand that incognito does not make you invisible and it does not enhance security. All it does is delete the tracks of your browsing when you exit. To be more precise, only the tracks *on your computer* are deleted.

Incognito has another nice side effect. Let's say you want to see how the rest of the world views your LinkedIn or Facebook page. Normally you'd have to log out of these sites to view them as any other person would see them on the web. If you use an incognito window, however, you can look at them as the general public even if you are still logged in on your non-incognito window.

All the larger search engines use your web surfing habits, among other things, to determine what to show at the top of the search lists. For example, if Google notices that you've been looking for a watch it may show a few links to watch shopping sites at the top of any searches for watches. Using an incognito window will let you search without these biases.

Finally, incognito is great for Christmas shopping. You can search to your heart's content without fear that your spouse or children will look at your computer and know what you've been searching for.

To create an incognito window open Chrome normally. This opens a non-incognitos window. Now click on the Chrome menu icon in the upper right-hand corner. Then select "New incognito window" from the menu.

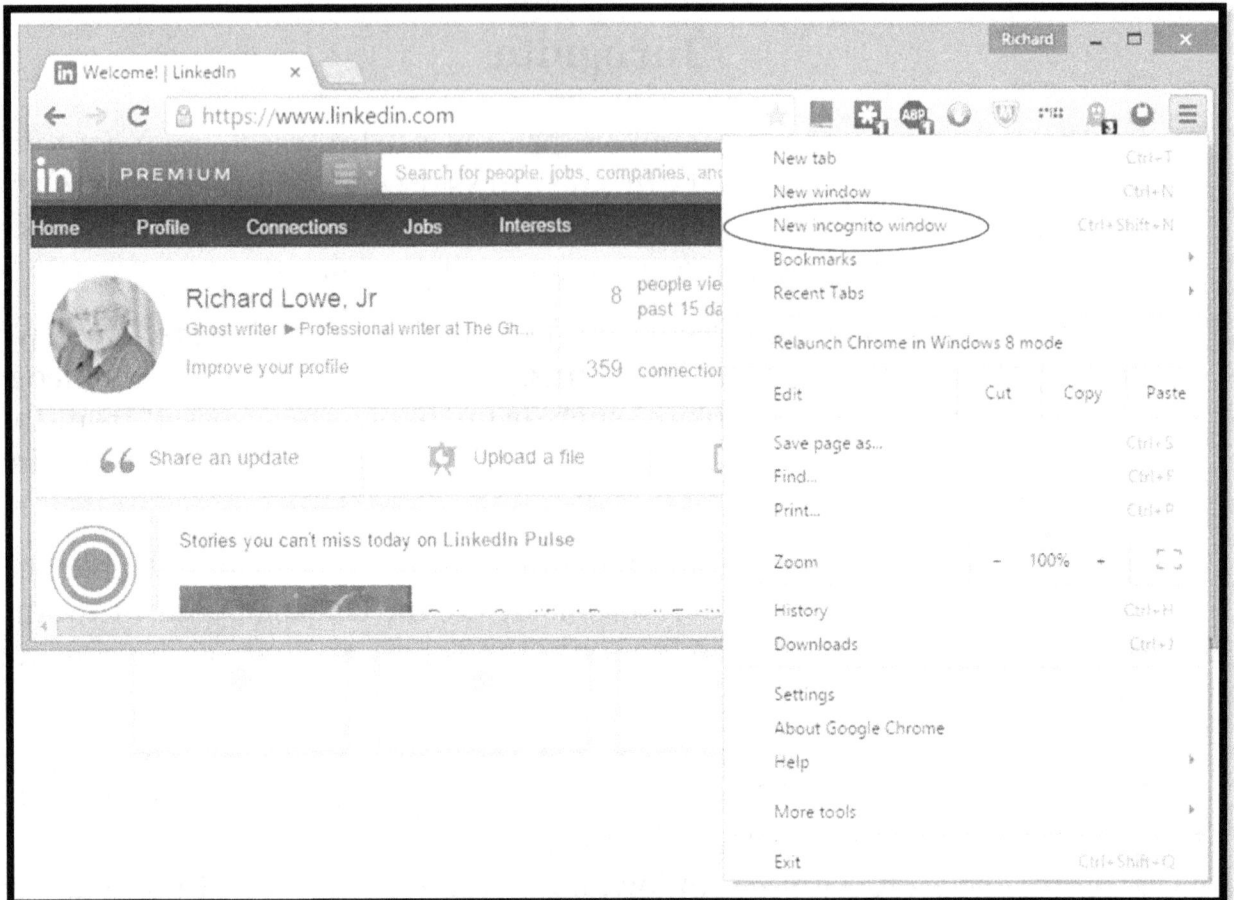

That's all there is to using incognito.

Safely downloading files from the Internet

Have you ever found a program you wanted for your computer on the Internet, but were afraid to download it because it might contain malware? So even though it was perfect for your task you passed it by. After all, who wants to install a virus on their computer? The program isn't *that* important.

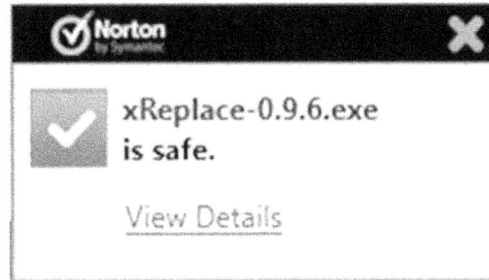

If you install Norton Security, whenever you download a file a small box will be displayed in the lower right hand corner of your screen. This box will tell you if Norton considers the file safe or malicious. You can click on View Details to find out more information.

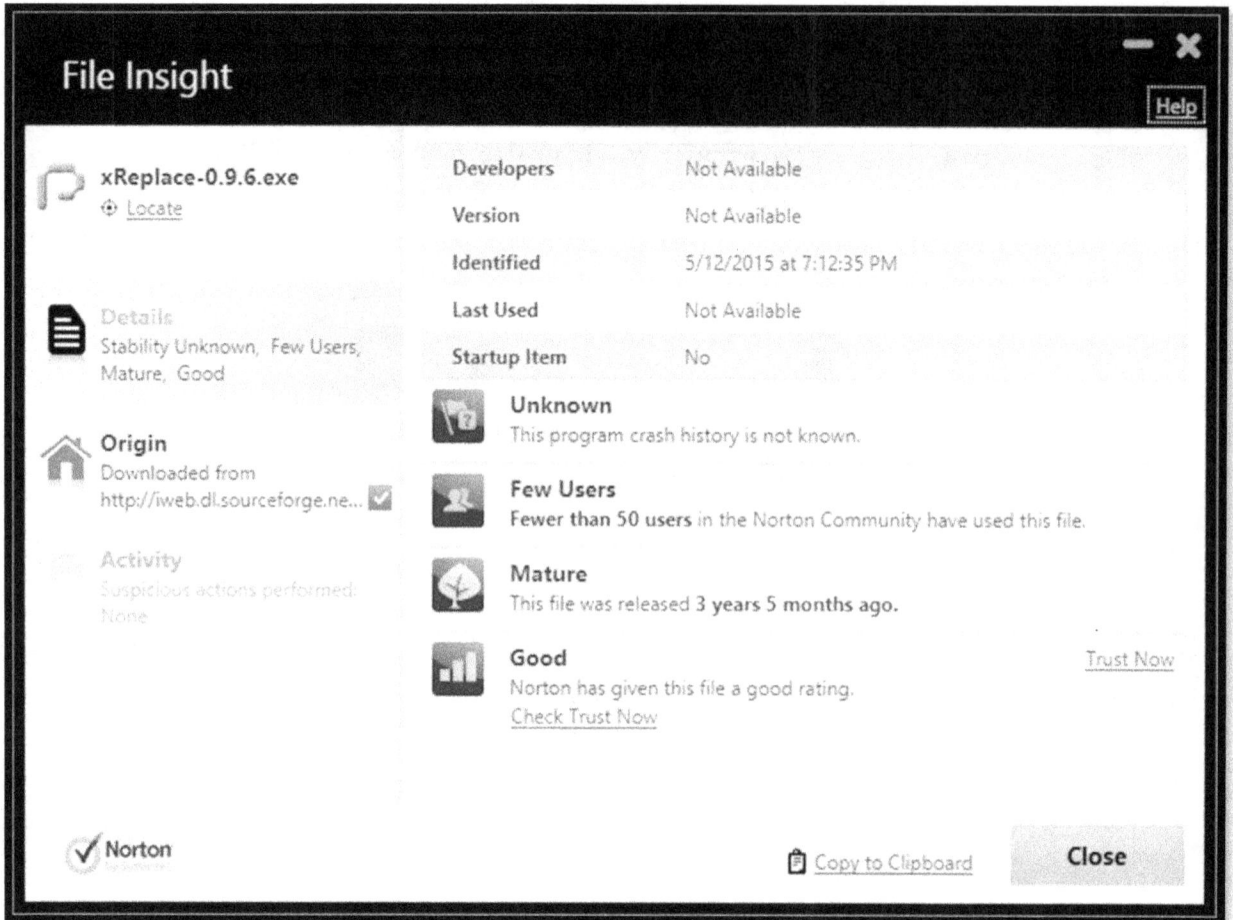

There are several online services which will scan a file. All of them will let you upload a file to be examined. Some of them will let you enter a web address so scan directly on the Internet.

Jotti's malware scan	http://virusscan.jotti.org/en
Metascan	https://www.metascan-online.com/
NoDistrubute	http://nodistribute.com/
ScanThis	http://scanthis.net/
ThreatExpert	http://www.threatexpert.com/submit.aspx
Virustotal	https://www.virustotal.com/

Once you've scanned a file you can have some certainty that it does not contain a virus.

However, you still need to review the terms and conditions of the product, also called a EULA, which is usually presented as part of the installation process. Quite often there is a checkbox where you acknowledge that you read and understand the terms. Most people just click the checkbox without reading the document.

Quickly read through the document. Look for verbiage which allows other applications to be installed. This is how some spyware gets installed, and it's completely legal since you authorized it by clicking the checkbox!

The site SpywareGuide has a nice tool into which you can paste the terms and conditions to be analyzed for suspect clauses. Read more about it here:

http://goo.gl/BDzJda

http://www.spywareguide.com/analyze/analyzer.php

Safe Searching

Have you ever been using your favorite search engine and clicked on a link only to find it leads to a pornographic or other undesirable site? I'm not making a judgment here about the ethics of gambling or other sites. Unfortunately, these are often the source of many computer infections.

The major search engines, including Google and Bing®, include a setting which filters out web sites that are not suitable for children. These settings are not perfect but they do screen out most of the explicit sites.

> **Best Practice**
> *Set filtering to maximum on your favorite search engines.*

The safe search setting for Google is on the following page:

http://goo.gl/dRhVs2

http://www.google.com/preferences

You can set Google to filter your results, and you can lock that setting with a password to keep someone (such as your children) from changing that setting.

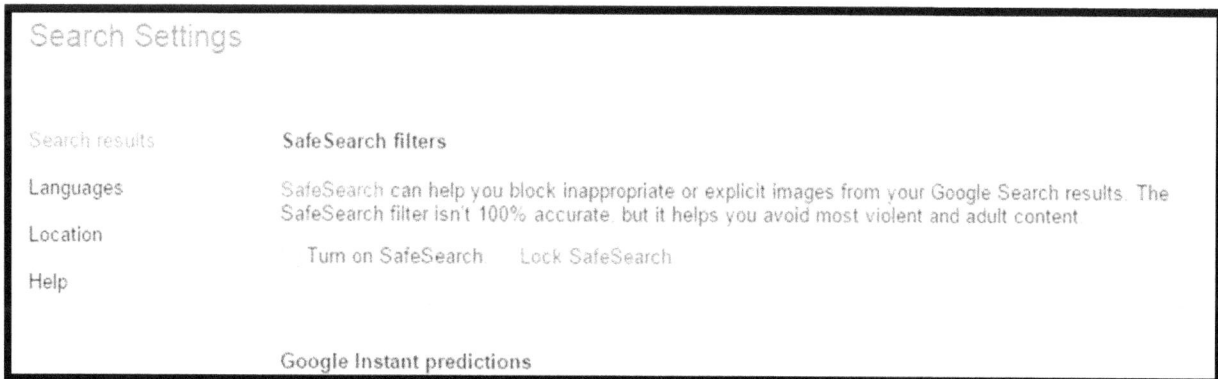

```
Search Settings

Search results        SafeSearch filters

Languages             SafeSearch can help you block inappropriate or explicit images from your Google Search results. The
                      SafeSearch filter isn't 100% accurate, but it helps you avoid most violent and adult content
Location
                         Turn on SafeSearch    Lock SafeSearch
Help


                      Google Instant predictions
```

In other search engines such as Bing and Ask®, you need to create an account. Once you do that you can go into their settings, find the filtering option, and set it the way you desire.

My recommendation is to set filtering to maximum unless you specifically want to receive unfiltered results. This allows you to have control over what the search engine displays and lets you choose when to view them and when not to view them.

Setting filtering settings can be a useful way to prevent you from accidentally clicking on a search engine result and going to an undesirable site. Unfortunately, there are hundreds, if not thousands, of search engines. Even if you could spend the time to create an account on each one, find the filtering option (if it even exists) and change it, doing so might have some security problems as well.

It is perfectly safe to create accounts and change settings on all of the major search engines. Unfortunately, there are some search engines which exist just to harvest information for spamming or other malicious purposes.

Thus my advice is to be very cautious about creating accounts and entering personal information in any search engine (or any other web site for that matter) which is not known to you.

Remove unneeded applications

If you are like most people you use your computer for a large variety of tasks. You probably have a word processor, a spreadsheet program, a graphics editor, and dozens of other applications that you've purchased and collected over the years.

It is very important that you remove applications that you no longer need or use. Every single application installed on your computer can affect performance, disk space usage, system stability, and security.

> **Best Practice**
> Remove unneeded applications on a regular basis.

In addition, applications require patching just like the operating system does. Since keeping applications up-to-date can be a major chore, usually requiring manual downloads and installs, it is often simply not done.

It is highly recommended that very once in a while you review your applications and uninstall those you no longer need or use. Use the Programs and Features control panel to perform the uninstall process on each application, one by one. Some applications will require a reboot after uninstalling.

Be sure to only uninstall those applications that you recognize. Some of the items in your Programs and Features control panel are required for your operating system and applications to work.

Uninstall Java

Java® is a set of tools which allow complex web sites to be created to do things such as play games or display pretty pictures. Some older web sites require Java to run, but the trend today is to eliminate it completely.

> **Best Practice**
> Uninstall JAVA from your computer.

Unfortunately, Java alone is a security risk. Unless you absolutely need it, you should uninstall it immediately. Don't worry, if a web site requires Java it will ask you to install it before it will run properly.

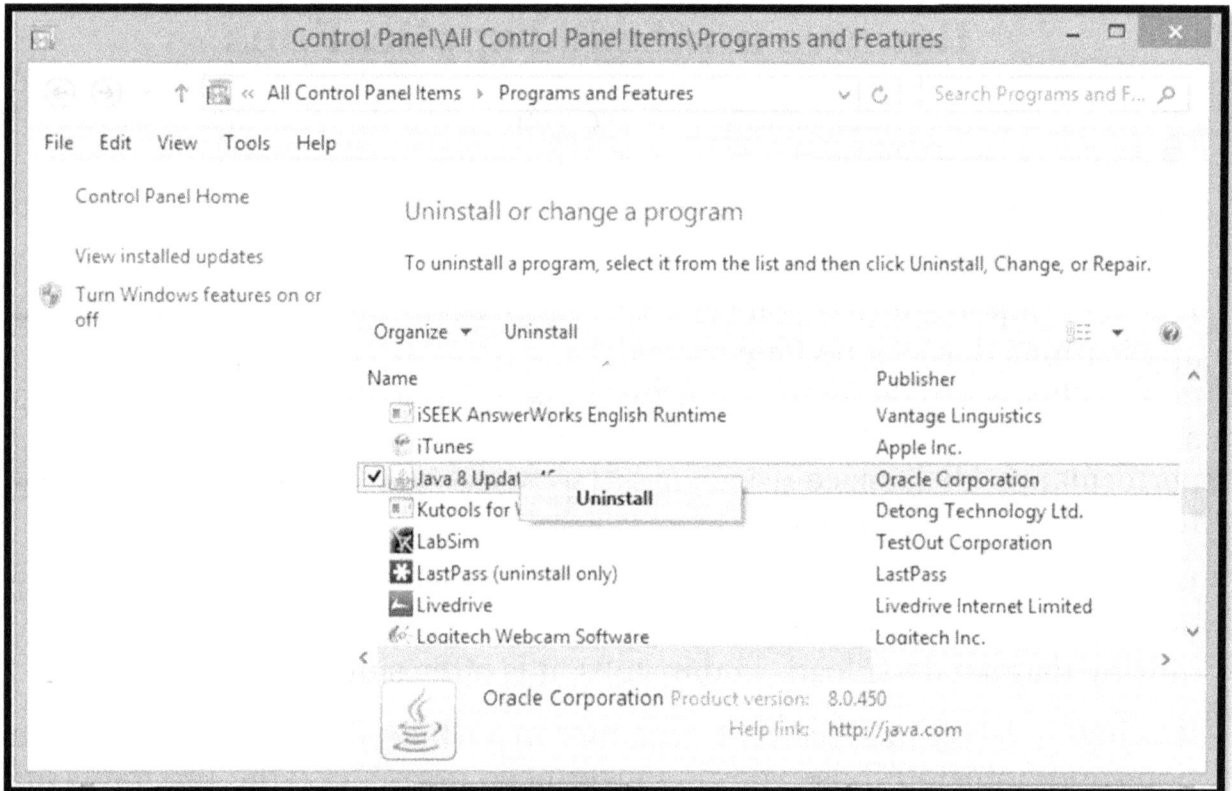

Be sure to uninstall all versions of Java that exist on your computer.

Remove all toolbars

> **Best Practice**
> Uninstall all toolbars from all of your browsers.

If you are like many people, you have at least one, and probably several, toolbars installed in your browser. To improve your security, the stability of your computer, and your systems performance, uninstall all of them.

Of course there are useful toolbars, so you will need to judge their importance to you.

Toolbars are installed directly into your browser. To remove them you need to look in several places.

- For Google Chrome, look under the Tools menu and selection Extensions. Remove the toolbars from there by clicking on the trash can icon next to the name of the toolbar. Note the toolbar may not have the word "toolbar" as part of the description.

- Some toolbars can be removed using the Programs and Features control panel. Just uninstall any that you find unless you have a good reason to keep them.

Toolbars are not necessary for the operation of your computer. Unless you have a pressing need for one, it is highly recommended to remove them all.

While uninstalling toolbars you may encounter some that will not allow themselves to be removed. Spyware and other malicious applications are not easily uninstalled. In these instances, you may need to consult with your local computer technician, antivirus vendor or some other knowledgeable expert.

There is no such thing as a free lunch

TANSTAAFL is a term from a book by Robert A. Heinlein (one of the best Science Fiction authors who ever lived) called *The Moon Is A Harsh Mistress*. The term means *There Ain't No Such Thing As A Free Lunch*. This concept is the basis of the book's plot, which is about a lunar penal colony and its attempt to free itself from Earth domination.

The term and its underlying concepts are important to remember while you are enjoying yourself on the Internet. There is so much free stuff available that sometimes it's easy to forget that very little of it is really free. After all, the bills need to be paid and everyone needs to eat.

Let me back up a minute and state that there are things that are free all over the Internet. Really and truly free. Some programmers write software and give it away as freeware. This is done for a variety of reasons, sometimes with noble sentiments and occasionally with more commercial desires, such as advertising their skills as programmers. Other people create web sites that contain information and entertainment which is freely available to everyone just because they want to help people or they enjoy writing.

However, when you are referring to corporations, there is virtually always a need to make money somehow. This should be very obvious, as very few employees will work for free, and most employers will not pay their employees out of their own pockets.

So how does, for example, Google actually make money? On the surface Google appears to be very noble, giving away tons of services and features for absolutely no cost. You can get your email, be part of groups, join a club, maintain a calendar, check movie listings, and perform literally hundreds of other tasks without sending them a dime.

What Google and other similar companies want from you is your demographics. They need to know how much money you make, what your interests are, where you live, your sex and age, and other similar information. Essentially they build up a packet of information about you, called a profile. This profile can be extraordinarily precise, including the sites you visit, purchases you've made (and for whom), physical locations you frequent, and so on. Every single thing you do on the Internet, including your personal computer, your tablet, your Kindle, and your phone, is recorded by some company somewhere.

This information is then boiled down into statistics and sold to advertisers. Let's say an advertiser is selling women's shoes, and the target audience is women over 40 and their husbands. The advertiser also knows from their own statistics that people who make between $50,000 and $75,000 tend to make larger purchases. They can tell this to Google, who will target the advertisement to precisely that market.

There are many other examples of TANSTAAFL throughout the Internet. In fact, you don't have to try very hard to find them, as one of the major business models of Internet commerce is giving free stuff to people in exchange for viewing advertisements.

So what's the point of all this? It's important to be aware of how you are actually paying for the service or product that you are receiving. That way you can make an intelligent decision as to whether you want to be part of the transaction. Otherwise, you are hostage to the company with which you are doing business.

A good place to begin is to read the terms and conditions of any services (paid or free) that you are using. You should reread them occasionally as they will change once in a while. These agreements are often complex and contain precise legal terms, so they can be difficult to confront and understand. Just grab a good dictionary and look them over.

Once you've read the terms and conditions, another good practice is to use Google or another search engine to find reviews and comments about the services you use. You will almost always find a few, as no service can please everyone. What you are looking for is not the occasional horrible review, as even the very best services get bad reviews sometimes. What you want to see is balance between good and bad comments and reviews. The content of these reviews is also important as a company may receive bad reviews on an offering that you do not use. Another excellent source of information is your local BBB web site.

A good place to look is in the promotional materials for stockholders. This is where a corporation explains to investors why they should be investing in their company. This often tells you exactly how the company makes money, and it can often produce some very eye-opening results.

Press releases are also good places to find information, as are privacy notices. Keep in mind that privacy notices can be changed, and there is some debate about their enforceability. These are statements of intent, not legally binding contracts.

I know this sounds like a lot of work, and you may not want to do it all for every single little free service that you sign up for. However, let's say you are using free web space from a provider. You put in lots of time and effort, only to find out that according to their terms and conditions they own the web site that you just created! Wouldn't that really annoy you, especially if they tried to enforce it?

Sometimes the terms and conditions produce pleasant surprises. I remember a huge outcry from many photographers who were under the impression that any photos they posted to Facebook became the property of Facebook. If these photographers had bothered to read the terms and conditions they would have known that the only right Facebook gained over their photos was a license to display them. The copyright remains with the creator of the photo.

The point is simple: rarely is there ever a free lunch that is honestly and completely free. It's a good idea to look around and be sure that the lunch does not have strings attached before you eat it.

How does all of this relate to home computer security? These companies maintain vast databases about you and every other person who uses the Internet. By informing yourself about who uses this information and what it is used for, you can make more intelligent decisions about your online practices. For example, if you don't want the large advertising companies to build up your profile, you'll want to install software that helps prevent them from doing so. On the other hand, if you don't care or even desire this information be maintained (it does produce more targeted advertisements) then you will want to use a different mix of applications.

Scams and hoaxes

There are countless hoaxes propagated all over the Internet. Every single day I see at least one or more posted to my Facebook feed, and a quick look at my spam folder shows dozens more.

A hoax is false information. Sometimes they are created just to annoy people, other times they are an attempt to discredit a person, political party, or company, and sometimes they are just malicious.

Some good sites to use to help you determine if something is a hoax are listed below.

Snopes	http://www.snopes.com/
Hoax Busters	http://www.hoaxbusters.org/
About.Com Urban Legends	http://urbanlegends.about.com/
Hoax Slayer	http://www.hoax-slayer.com/

It is generally a good idea to take a quick look at these sites before furthering a hoax.

The Nigerian scam

Have you ever received an email which read like a soap opera and claimed some king or famous person had left a massive amount of money in a bank in the United States? Did you see how the sender of the message explained that he wants your help to retrieve that money, and offers you a large percentage as a commission? There are as many variants of this as there are stars in the sky, but they all prey on the gullible, the greedy, and the desperate.

This is often called the Nigerian scam, the 419 scam (419 is the penal code in Nigeria for advanced fee fraud) or the advanced-fee scam. In spite of the name, these scams originate from all over the world, although a very high percentage of them do originate in Nigeria.

Here is an actual example which I plucked out of my scam message folder.

> *Good Day*
>
> *I saw your contact email from Nigerian Chambers Of Commerce and decided to contact you after much prayers and consideration since I cannot be able to see you face to face for now, at first I strongly believed that any information received from the Chambers Of Commerce Office is correct and must be relied on.*

How are you today? I know that this mail may come to you almost a surprise as we never met before, but before you proceed reading this mail, I want you to settle your mind and read with good understanding of the situation, this is true and not a joke. However,

I am Barrister davidmadu, Attorney to the late Engr. Steve Moore, a national of Northern American, who used to work with Shell Petroleum Development Company (SPDC) in Nigeria, on the 11th of November, 2004. My client, his wife and their three children were involved in a car accident along Sagamu /Lago Express Road.

Unfortunately they all lost their lives in the event of the accident, since then I have made several inquiries to several Embassies to locate any of my client's extended relatives, this has also proved unsuccessful.

After these several unsuccessful attempts, I decided to trace his relatives over the Internet to locate any member of his family but to no avail, hence I contacted you, I contacted you to assist in repatriating the money and property left behind by my client, I can easily convince the bank with my legal practice that you are the only surviving relation of my client, otherwise the Estate he left behind will be confiscated or declared unserviceable by the bank where this huge deposits were lodged. Particularly, the Bank where the deceased had an account valued at about $15 million U.S dollars (Fifteen million U.S. America dollars), Consequently, the bank issued me a notice to provide the next of kin or have the account confiscated within the next ten official working days.

Since I have been unsuccessful in locating the relatives for over several years now. I seek your consent to present you as the next of kin to the deceased, so that the proceeds of this account valued at $15million U.S dollars can be paid to your account and then you and I can share the money. 50% to me and 50% to you.

All I require is your honest cooperation to enable us see this deal through and also forward the following to me so that I can file an application of claim in your name to the bank immediately:

1, Your Full Names: ....

2, House Address:...

3, Your Country: ....

4, Your Contact Telephone Number:......................

5, Your Age and Gender:...

6, Your Occupation:...

I guarantee that this will be executed under a legitimate arrangement that will protect you from any breach of the law.

Please get in touch with me VIA my confidential email address as follows

<censored>@gmail.com

Barrister davidmadu.

As you can see from the example, this message is intended to hook you into the scam by promising great riches for virtually no risk. The back story, often with soap opera-like plots, lends credibility to the narrative, and ropes you in. If you respond to the message the scammer will present various documents and photos to attempt to prove the story is true.

Once your confidence has been gained, a roadblock of some kind will appear. This might be a bribe that is required or money that needs to be deposited in some bank out of the country. The scammer will almost always ask the money be sent using a wire transfer; this is because wire transfers are untraceable. The scammer will take your money and keep it for themselves.

Now the scammer will tell you there is another roadblock to overcome, then another one and then another until you run out of patience or money. The fact that mildly illegal acts such as bribery were involved is designed to prevent you from going to the authorities. In order to report the crime you'd have to confess to committing a crime.

Once the scammer is sure you've run out of money or will not send more, they either take the money and run (if you are lucky) or they invite you to some location outside the country. If you do agree to meet them, once you arrive you will be robbed, probably beaten, and possibly held for ransom. There are reports that some individuals have been murdered.

There are several variations on this scam.

- Lottery scams – These involve receiving fake notices of winning a lottery. If you respond the scammer tells you he will send the money if you send him a small fee. If you do pay, the scammer will invent another fee and then another until you can no longer pay.
- Employment scams – If you've posted a resume to a job site you may have received one of these messages. The scammer sends you a fake job offer out of the country with an excellent salary and benefits. Once you reply they will ask for processing fees. Of course the offer is fake, the money is lost, and the job is non-existent.
- Various check cashing scams – There are many variants of this scam. Basically they involved the scammer sending a check to you, requesting that you cash the check, subtract a commission, and wire the cash back to them. Of course the check bounces and you lose the money you sent.

- Online sales scams – In these scams the criminal purchases the goods with a check made out for a larger amount than the purchase. The idea is to get you to send the difference to them in cash. After a few weeks the check will bounce and the money will be lost.
- Romance scam – This scam is common on Internet dating sites. The scammer will contact you, gain your confidence, and get you interested, but needs money to get into the country. The scammer explains he or she needs to buy a plane ticket, rent a room, and so on and asks you to wire them the money. Of course the money is lost.
- Assistance scam – I've seen this one on Facebook. A friend's account was hacked (she had a terribly weak password) and the scammer pretended she was my friend and sent me a chat request. She tearfully claimed she was stuck in England, falsely arrested, and needed a couple of thousand dollars to get bailed out. Because I had just been to a show with my friend a few hours before I knew she wasn't in England.
- Reloading scam – This is an interesting scam. The scammer contacts a previous victim of one of the scams and claims they can help track down the money. Of course they need a fee for this service, naturally in advance, and via wire transfer.
- Other scams – "Collections" companies looking for you as well as the IRS saying you owe money.

It is best to simply delete these messages upon receipt. Better still, any good spam filtering system should delete virtually all of them before the messages get to your inbox.

Chain messages

A chain email is a message that asks, or demands, that the recipient send the message along to others. Chain messages at best needlessly use resources as they're sent to thousands or millions of people, and at worst are illegal.

One famous snail mail variant claimed you can make thousands of dollars with a six dollar investment (plus the price of 6 stamps.) All you need to do is send a dollar to everyone on the list, add your name to the top of list, remove the top name from the list, and send it out. This is what is called a pyramid scheme and it is illegal.

Sometimes these chain message demand you resend them to a certain number of people for good luck, to help some cause, or just for kicks.

Fortunately, the spam filters of most email systems will catch these chain messages so you don't see them. If they do get into your inbox simply delete them.

Device Security

D id you know that in addition to the Internet, the web and email, your computer can also be infected by devices and media? Viruses can be spread by SD cards, flash drives, CDs, DVDs, and literally anything else connected to your system.

Not only that, hackers who gain control of your system also can control your cameras and microphones. They can turn on the camera remotely and watch anything in view of the lens, as well as listen on the microphone.

Autoplay

Autoplay is a feature of Windows which causes an application to be run when media (DVD, CD, flash drive, etc.) is detected.

On older versions of Windows called this AutoRun, and it was a huge security problem. In the past, when you inserted media into your computer Windows would look for a file called autorun.inf and do what that file instructed it to do. This meant that, for example, a floppy disk could contain an autorun.inf file that instructed Windows to install a virus.

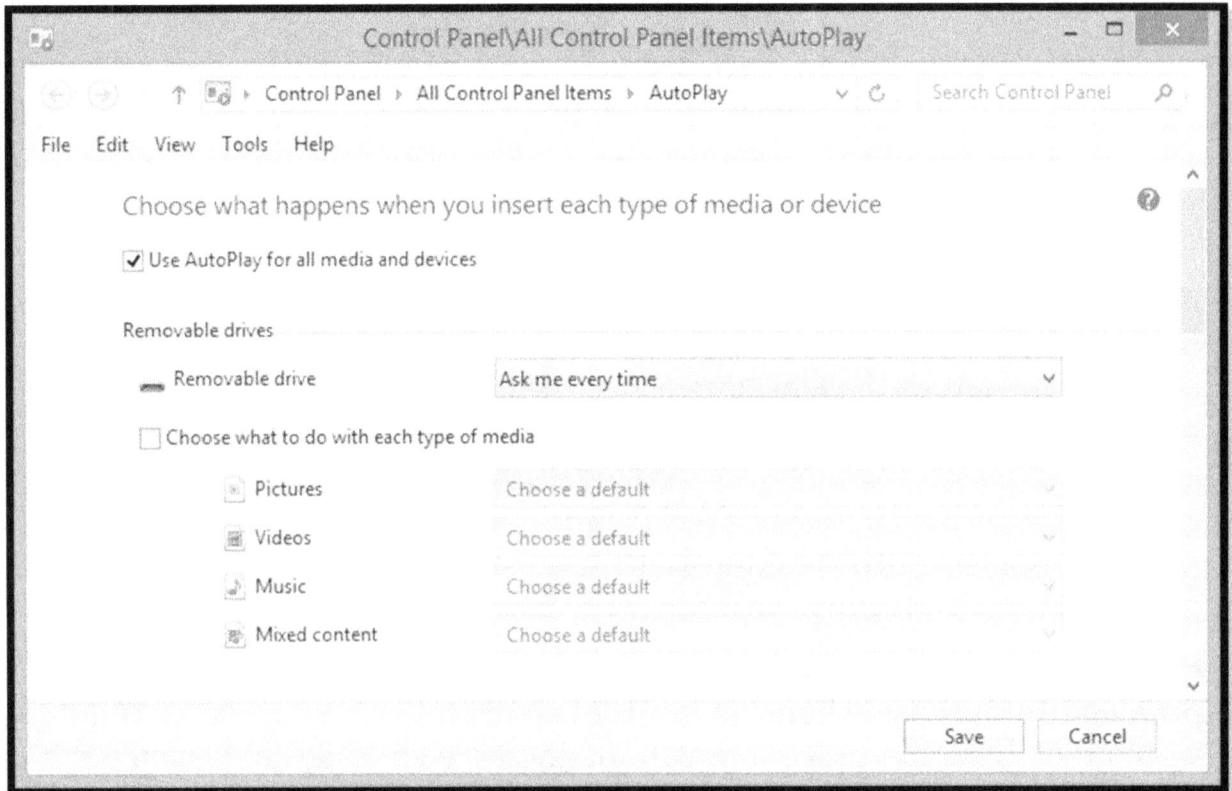

AutoPlay modified this behavior to examine the media and present a list of possible actions for you to take. Thus a CD would ask you to play the songs, copy them to your music library, open up the media in Windows Explorer, or do nothing. If an autorun.inf file is found, it is added to the list of options. This prevents autorun.inf from being automatically run when media is inserted.

From a security standpoint, the safest course of action is to disable AutoPlay entirely. Just remove the check the "Use AutoPlay for all media and devices" in the AutoPlay control panel.

SD cards

SD cards, or any card that is similar, are often used in digital cameras as storage for photos. Most modern computers include at least a slot for an SD card, and many include several slots for all the different card formats.

The primary advantage to SD cards is they are small and inexpensive. This makes them the perfect storage media for cameras and similar devices. Even smaller cards, called Micro SD cards, are used in smart phones. They can be accessed by a computer if you purchase a special conversion card.

Any SD card can include a virus which can infect your system when a card is inserted into your computer. Used cards should be considered suspect, and any SD card found in the trash or on the ground should be destroyed.

The other problem with SD cards, as with any device, is data that your store on them can be accessed even if you format them. It is quite common for people to find personal photos on used SD cards. Before discarding any SD or other similar cards you should either erase them or completely destroy them.

Remember to remove the Micro SD card before you get rid of your phone. That's because the data on that card can be retrieved by whoever gets the phone.

Flash drives

Have you ever found one of those little flash drives laying on the ground? Were you tempted to pick it up and use it for your own data? Don't do it! Even though it seems like a waste, leave the flash drive on the ground or in the trash.

It is impossible to know what has been put on these discarded drives. They could contain viruses or other malicious programs specifically put there to attack computer systems.

Some hackers have been known to purchase flash drives by the case, install malicious applications on them, and scatter them were they will be found. These small drives are almost irresistible because they seem valuable. The same problem can occur with cards purchased used.

Unfortunately, even flash drives that you purchase from a store can be compromised. It has been reported that some flash drives have been infected with malware at the factory.

How do you protect yourself from this threat?

- Discard any flash drives which you find on the ground, in the trash, or anywhere else.
- Ensure your computer has been secured as discussed in this book.
- Don't purchase used flash drives (you won't save that much money if you do, anyway).

Be sure to destroy any cards that you discard. Even if you format the cards the data can still be recovered.

Web cam security

I'm sure you know that virtually every tablet and laptop made these days comes with at least one, and sometimes two, built-in cameras. These cameras can be used to take still photos and video, and they are great for video chatting. For larger computers, such as a desktop system, you can purchase cameras which plug into the USB port.

If your computer is compromised by malicious software, such as a virus, then it becomes possible for a hacker to gain access to your webcam. Once this happens they can turn it on and view everything within range of the lens. Some cameras can even be turned on without the LED being lit up.

> **Best Practice**
> Cover the lens of your computer's webcam when you are not using it.

The situation is even worse with webcams that do not require a computer. These are the webcams which connect to the Internet directly. The problem with these stand-alone devices is they are generally not updated with security fixes (patches) and thus become insecure over time.

What can you do to keep your webcam secure? Follow the recommendations in this book to secure your computer. If you are really concerned with the security of your webcam, cover it up with a piece of paper or something when you are not using it.

Wrapping it all up

In the past, operating system and computer security was an afterthought. What came first were the features that attracted people to make a purchase. In other words, marketing came before security. That mindset has, for the most part, changed. Modern computers generally include reasonably strong security straight out of the box.

This does not mean you can ignore security. All operating systems have gaps and vulnerabilities that hackers can use to gain access. These gaps can be filled by the judicious use of several security applications.

More importantly, the single most critical part of any security plan for your computer is *you*. The tasks that you perform every day can either hinder or help your security. By following the best practices defined in this book you can reduce the chances that your computer will be comprised.

The threats are real and they change every single day. Hackers find new ways to break through your computer's defenses, and vendors such as Microsoft provide updates and services to fight off those attacks.

There is war going on as we speak. Malicious individuals and groups have a strong desire to break into your computer for many reasons. Sometimes they want to steal information, money, or property. Often they want to siphon off your computer resources by adding your system to a botnet. Occasionally, these people will try to steal your identity, and, in the case of the Nigerian scam, it is possible they may try to take your life.

Information security issues are getting hotter every day. Viruses and other malicious applications are being created by huge criminal organizations based all over the world. To make matters even worse, major countries such as the United States, Israel, Russia, and China are creating viruses for the purposes of warfare, and this is being called cyber warfare.

I am hoping that by having read this book and understanding the presented concepts that you can make your computing environment safer. In turn, this can save you an immense amount of money, heartache, and grief.

Glossary

This glossary is intended to give you a short, non-technical understanding of the words. The definitions are often simplifications and specific to the needs to this book.

Account

> When someone logs into a computer they are working in an account. An account defines what can be done on that computer.

Administrator account

> A windows account which can perform installations, patching, and other tasks which normal user accounts cannot do. No user should ever use the administrator account for day-to-day activities. Log into individual user accounts (one for each person) and only use administrator when needed.

Antivirus

> An application designed to scan the disks of a computer for viruses, and remove those viruses once found.

Applications

> A computer program created to perform a task.

Autoplay

> This setting causes an action to happen when media (DVD, CD-ROM, SD card, and so on) is detected or inserted. Each device can be set to play music, video, open a folder, and other similar functions.

Autorun

> See Autoplay.

Back Door

> A secret way to login to a system.

Banners

> Advertisements displayed on web pages.

Best Practices

Generally accepted recommendations of how to best secure your system from various types of attacks.

Black hat

Hackers who engage in criminal activities.

Bot

A virus designed to take over a computer to allow a hacker to remotely control it for their own purposes.

Botnet

A network of computers who have been taken over by a hacker or criminal organization. Some botnets are comprised of millions of computer systems.

Browser

An application such as Google Chrome, Internet Explorer, Apple Safari or Opera which is used to interact with web sites on the Internet.

Browsing

The act of using a browser to explore the web. These browsers display pages and allow users to click links to move from site to site.

Chain email

A message that asks, or demands, that the recipient send the message along to others.

Clickbait

Headlines which are designed to get someone to click. They are usually untrue and are just a way to entice people to view a web page or a video.

Cloud

Resources which are out on the Internet instead of being located in a home or office.

Computer terminal

A screen and keyboard connected to a larger computer system.

Connect

To successfully establish a method of communication. For example, when connecting to a web site, the browser displays the web site in a window.

Control panel

A utility on Windows computers which allows various system features to be manipulated.

Cookies

A cookie is a small text file stored on your computer to identify you to web servers.

Cracker

A person who breaks into computers.

Cyberwar

When nations use computers to attack the computer resources of each other. The Stuxnet virus, for example, was created by the United States and Israel to attack the computers controlling the motors of the centrifuges used by the Iranians to create nuclear fuel.

Data Execution Prevention

A security feature to help keep a computer more secure.

Denial of service attack

An attack whose purpose is to make a web site or network unresponsive.

Device

A piece of equipment which, for the purposes of this book, is connected to a computer or a network. A desktop computer is a device, as are digital cameras and printers. A smartphone (regardless of the model), tablet, video game consoles, scanners, and even smart light bulbs are also devices. Any computer, anything that can connect to a computer or anything attached (or that can attached) to a network is a device.

DHCP

> Dynamic Host Configuration Protocol. The method of assigning internal IP addresses to your computer. DHCP also refers to the application that performs this function.

Distributed denial of service attack

> An attack from a large number of computers, typically a botnet.

DNS

> Domain Name System. The function which translates names such as google.com into IP addresses.

Domain Name

> A name such as google.com, microsoft.com, or leave-me-alone.com. DNS allows web sites to be identified using names instead of IP addresses.

Domain Name Service

> See DNS.

Download

> Copying data from the Internet or another computer to your computer.

Drive-by

> Visiting a web page which secretly infects a computer with a virus. The user does not need to do anything except view the page.

DSL

> Digital Subscriber Line. A means to connect to the Internet using phone lines.

Dumpster Diving

> Physically looking through someone's trash to find passwords, credit card numbers, and other information.

Enhanced Mitigation Experience Toolkit

> A tool which you can install that prevents fill many of the gaps in security on a Windows system.

EULA

End-User License Agreement. The legal contract defining how the product may be used.

FIOs

Fiber Optics Service. A method to connect to the Internet using optical cables. These provide for extremely fast uploads and downloads.

Filtering

To screen out certain web sites based upon content, name of the site, and other criteria.

Firewall

Literally a wall, albeit electronic, between a network and the Internet; it keeps the bad guys out.

Firmware

Programs which run directly inside a device to make that device function. Thus cameras, printers, disk drives, and every other device has its own special firmware which actually operates the device itself. Each personal computer has its own firmware to control the computer hardware.

Flash Drive

A small electronic device used to store data and transport it to another computer or camera.

FTP

File Transfer Protocol. A method for moving files around the Internet

Guest network

A network you can create on your own router which is for the use of guests. This lets visitors to a home get on the Internet without getting into your network or using the main passphrase.

Hacker

According to the Merriam-Webster online dictionary, a hacker is defined as "a person who secretly gets access to a computer system in order to get information, cause damage, etc.: a person who hacks into a computer system."

Hacker Underground

Generally, computer users, web sites, and so on that engages in illegal or illicit activities. These include virus creation, phishing, warez, and pirating, among other things.

Hardware

Computers and other physical devices.

Hoax

An article or email which intends to deceive.

Hotspot

A wireless connection to the Internet. A computer can connect to the web using hotspots in coffee shops, train stations, and so on.

HTTPS

HyperText Transfer Protocol over SSL (Secure Sockets Layer). A part of the web site address which indicate the connection is secure. This enforces security for the connection.

Incognito

A function of a web browser which allows you to surf the web without leaving tracks on your own computer.

Internet

A simplistic definition is a network composed of all of the networks in the world. This is not strictly true as many networks are not connected to the Internet, but the definition is useful for this book.

Internet Provider

A company, such as Verizon, that provides access to the Internet to homes and businesses.

IP Address

An address used to identify a connection to a network.

Internal IP Address

See NAT address.

Java

A set of tools used to create applications for use on the Internet.

Key logger

An application which records every single key that is typed and every mouse click that occurs on a computer. These are often planted by viruses in order to get passwords and other information.

LAN

Local Area Network. The network within a home or office.

Link

Some text or a graphic which can be clicked with the mouse to cause the browser to display a different web page.

Logic bomb

This is a virus or piece of code, often installed by hostile employees or contractors, which is timed to trigger its payload at a specific date in the future. For example, a malicious employee might leave some code embedded within the accounting package which causes it to delete everything a year in the future. These types of infections are very difficult to detect and even harder to eradicate, as they may have been added to the system years before. Even backups may be corrupted with the malicious code, so restoring may not be an option for recovery.

LoJack

A service which you can purchase which allows a computer to be located in the event it is stolen.

MAC Address

A number used to identify a device on a network.

Malware

This stands for malicious software. It means anything which intends to do a person or their computer system harm, or to a computer to harm other computer systems or people.

Man-in-the-middle attack

Listening in on the communications between a computer and a web site. This is often done by hackers who create hotspots they control.

NAT Address

Network Address Translation. The address assigned to a computer inside a local network. These address are in the form 192.168.n.m where n.m are two different unique numbers.

Network

Two or more computers that are connected in order to share things (devices and resources).

Operating system

The application that administrates (operates) a computer. Windows is the Microsoft operating system and it works on PCs, tablets, and other devices. Android is the operating system created by Google, IOS is for Apple iPhones and other devices, and the MAC OS X is for Apple computers. There are many other operating systems, but the average user is unlikely to be using them on their home computer.

Password

Characters entered into a web site, application, or computer required to gain access.

Patches

Updates to applications or the operating systems to correct errors.

Patch Tuesday

The day Microsoft patches are release to the general public, which is the second Tuesday of every month.

Payload

This is the part of a virus which performs the intended action. This could include adding the computer to a botnet, recording keystrokes, or destroying data, among other things.

Phishing

Emails or web pages pretending to be an official web site in order to steal credentials.

Pirating

See Warez.

Programs

A list of instructions which tell a computer what to do. Thus a spreadsheet program (also called an application) is the commands specifying to the computer how to manipulate spreadsheets.

Ransomware

Some of the more malicious viruses actually hold computer systems for ransom. These often first appear as a popup claiming a system is infected. They then ask to install "antivirus" software. Clicking anything installs is a very-hard-to-delete virus. This virus demands that a ransom of a certain amount (usually a few hundred dollars) for the safe return of the files. Some of the more sophisticated varieties actually encrypt files so they cannot even be used until ransom is paid. Often the only way to recover from these types of viruses is to restore files from a backup.

Resource

Any device or thing that can be used by a computer to serve some function. A disk and a flash card are resources. In addition, a shared folder (made available to others) is a resource. Printers, scanners, game consoles, and similar devices are also resources.

Restore Point

A point-in-time snapshot of the current operating system state. This allows the operating system and applications to be restored to the way they were at that time.

Router

A switchboard of sorts that connects different networks together. For example, the DSL modem or router connects a home network to the Internet. Home computers use the router to find out the best way to connect to web sites on the Internet.

Router IP address

The address of the router on the web. Used to identify the router on the Internet.

Scam

See hoax.

SD Card

A small card inserted into cameras and computers to provide data storage.

Server

A computer that serves a purpose or function. When playing games over the Internet using the Xbox, the game console is communicating to a server (actually many servers in what is called a **server farm**) which provides gaming services. There are many millions of servers doing a huge variety of tasks all over the world.

Server Farm

A group of servers in one location. Typically multiple servers act as one.

Shoulder Surfing

Looking over someone's shoulder while they are entering passwords or PINs.

Snatch and grab attacks

Literally when a thief grabs something valuable and runs away with it.

Sniffing

Intercepting wireless signals in order to listen in on the communications between a computer and a web site.

Social engineering

Tricking people into bypassing or breaking normal security procedures, or, more simply, conning them into doing something having to do with computer systems.

Software

Computer programs designed to perform various functions. Spreadsheets and word processors are examples of software.

Spam

Messages sent indiscriminately to many people at the same time.

Spamming

The act of sending spam.

Spear Phishing

This specialized form of phishing occurs when a specific company or group is targeted. Standard phishing tends to be sent out to a random group of people, while spear phishing is targeted. These attacks can be very specific, with messages discussing relatives, co-workers, competitors, and so on.

Spyware

This is application or technology which gathers information via computer without consent. Although spyware is unethical it is often legal because users themselves authorize it to be installed on their system. Keep in mind that the terms and conditions of many applications frequently authorize spyware to be installed.

SSID

Service Set Identifier. A name used to identify a wireless hotspot on the network.

Surf

See Web Surfing.

Time Bomb

This is a virus which triggers at some date in the future.

Toolbars

An application installed into a web browser, especially Internet Explorer, to add extra features.

Trojan horse

This type of malicious code is a virus or other dangerous program which is embedded within a desirable application. For example, someone might post a very nice screen saver on their web site which includes a Trojan horse that deletes files, sends information to a criminal origination, or perhaps just waits for instructions (this is called a bot or zombie).

Two factor authentication

In addition to a username and password, some other data is required to log into a web site or computer. Many web sites are sending a code to a phone which is then used to finish the log in.

Uninterruptable Power Supply

A large battery which acts as a surge protector and has enough power to keep a computer and some devices running for a few minutes in the event of a power failure.

Universal security slot

This is a small slit, about half an inch long (half the size of a USB port) located on the side or back of your laptop. The slot often has a "lock" symbol next to it. A security cable can be inserted into this slot to tie the system to furniture to prevent snatch and grab attacks.

Upgrade

To install a newer version of software on top of an older version.

Upload

Copying data from your computer to another system.

UPS

See Uninterruptable Power Supply.

Username

A name which is used to identify a user or person. On many web sites, and in Windows 8 and above, the username is the email address.

Utility

A special kind of application intended to do something technical for your computer.

Version

Different releases of programs, applications, or operating systems.

Virtual Private Network

A secure connection over the Internet or within a network.

Virus

A virus is defined as an application which can create copies of itself and install those copies on computer systems. Generally a virus tries to remain hidden so it can do what its creator intended without interruption from the user of the computer. Usually a virus consists of two parts. First is the delivery system, an email or web page or something similar, which gets the virus onto a computer system. Once the computer has been penetrated, the virus will install a payload, which is the actual virus application.

WAN

Wide Area Network. A large network. In many instances this is used to refer to the Internet.

Warez

Versions of applications such as Microsoft Office or Photoshop which have been hacked so they no longer need a license to be used. This is also called Pirating, and it is illegal. Warez applications are often made available for download on hacker web sites.

Web

The part of the Internet which is accessed using web browsers such as Internet Explorer, Google Chrome, Firefox, Opera, or Apple Safari.

Web Bugs

A very tiny graphic hidden on web pages to track online actions and user activity.

Web Cam

A video camera often attached to a computer which transmits video to a web site. Some models of web cams do not require a computer and transmit to a web site directly.

Web page

A document on the web which can be displayed by a web browser.

Web Surfing

Using a web browser such as Google Chrome to navigate between web pages and sites.

White Hat

Hackers who use their hacking skills to help improve security.

Windows

The most popular desktop operating system.

Windows Update

An application which downloads and installs security and other corrections, only from Microsoft, on your Windows computer system.

Wireless

Networks which allow systems to connect without wires using electromagnetic signals.

Wireless Hotspot

See Hotspot.

Wireless Isolation

A setting in wireless networks which prevents systems from access the Internet. In other words, they can only access the Internet.

World Wide Web

See Web.

Worm

A worm is a self-replicating virus. Some of the more common worms, such as "I Love You," use Outlook to send themselves to every email address listed in the contact list. Others actually have their own email system built into themselves so they can send to every email address they can find in any file on the hard drive. Some worms, such as Nimda, actually install themselves on web servers and then search through the Internet for other vulnerable machines. When these machines are found, the worm penetrates them and installs itself automatically.

WPA2-PSK (AES)

Wi-Fi Protected Access (WPA); Phase-Shift Keying (PSK); Advanced Encryption Standard (AES). A setting in wireless networks which ensures secure connections. This is generally the most secure method.

Zombie

See Bot.

Index

Recommended Reading

Data and Goliath: The Hidden Battles to Collect Your Data and Control Your World

By Bruce Schneier

An eye opening book detailing how information about you and everyone else is being collected. This author discusses how the data is gathered and used. Don't read this book unless you are ready to confront the information gathered about you and the uses to which it is put.

Dirty Wars: The World is a Battlefield

By Jeremy Scahill

An investigative report on some of the covert operations executed by the United States in the war on terror. The story of the Stuxnet virus, how it was created, and what it did to the Iranian nuclear program is fascinating reading.

Hacking: Ultimate Hacking for Beginners, How to Hack

By Andrew Mckinnon

Everything you wanted to know about hacking and how to hack a system. The book goes into detail on each of the major tools hackers use as well as the different types of hacks.

Incognito Toolkit - Tools, Apps, and Creative Methods for Remaining Anonymous, Private, and Secure While Communicating, Publishing, Buying, and Researching Online

By Rob Robideau

A basic book which describes how to shield yourself from prying eyes while you are online. It contains useful information which can be implemented easily to more or less cloak your information.

Social Engineering: The Art of Human Hacking

By Christopher Hadnady

If you want to know about social engineering and how it works, then this is the book for you to read. This is the art of mind tricks, how to convince someone to hand over the secrets or give you the keys. To make them relax and trust you so you can steal their secrets, their money, or their identity.

Spam Nation: The Inside Story of Organized Cybercrime-from Global Epidemic to Your Front Door

By Brian Krebs

An investigation into the criminals who are responsible for the global epidemic of cybercrime. If you want to know the story behind spyware, viruses, fraud, phishing, and other hacking techniques, then you will want to read this book.

About The Author

https://www.linkedin.com/in/richardlowejr

Richard Lowe has leveraged more than 35 years of experience as a senior computer manager and designer at four companies into that of an author, blogger, ghost writer, and public speaker. He has written hundreds of articles for blogs and ghost written more than a dozen books. He's published factual books about computers, the Internet, surviving disasters, management, and human rights. He's currently working on a ten-volume science fiction series, to be published at the rate of three volumes per year beginning in 2016.

Richard began in the field of Information Technology, first as the Vice President of Consulting at Software Techniques, Inc. Because he craved action, after six years he moved on to work at two companies at the same time: - he was a Vice President at Beck Computer Systems and the Senior Designer at BIF Accutel. In January 1994, he found a home at Trader Joe's as the Director of Technical Services and Computer Operations. He remained at that wondrous company for almost 20 years, before taking an early retirement to begin a whole new life as a professional writer. He is currently the CEO of The Writing King, a company which provides all forms of writing services, and the Senior Writer, Business Division, for The Ghost Publishing.

Richard has a quirky sense of humor and has found that life is full of joy and wonder. As he puts it, "this little ball of rock, mud, and water known as Earth is an incredible place, with many secrets to discover. Our corner of the universe is filled with beings, some happy and some sad, who each have their own special story to tell."

His philosophy is to take life with a light heart; He approaches each day as a new source of happiness. Evil is ignored, discarded or defeated; Good gets helped, enriched and fulfilled.

Richard spent many happy days hiking in many national parks, crawling over boulders, and peering at Indian pictographs. He toured the Channel Islands off Santa Barbara, and stared in fascination at wasps building their homes in Anza-Borrego. Some of his joys include photography, and he has photographed more than 1,200 belly dance events as well as dozens of renaissance fairs all over the country.

Because writing is his passion, Richard remains extremely creative and prolific; each day he completes between 5,000 and 10,000 words, diligently using language to bring the world to life so that others may learn and be entertained.

Richard is the CEO of The Writing King, which specializing in fulfilling any writing needs. You can find out more at http://www.thewritingking.com and emails are welcome at rich@thewritingking.com

Richard is a Senior Writer, Business Division at http://theghostpublishing.com

Colophon

All text written by Richard G. Lowe Jr.

Entire contents Copyright © 2015 by Richard G Lowe Jr. All rights reserved.

This book was written using Microsoft Word 2013. Screenshots were created for this book using Snagit by TechSmith. Each screenshot was edited using Paint Shop Pro X7. The font used throughout the manuscript is Georgia.

Cartoon graphics were all created specifically for this book by Retrosleep. You can find him at https://www.fiverr.com/retrosleep

The front cover was designed and created by Crownzgraphics. Their email address is crownzgraphics@gmail.com. The back cover was created by Richard G. Lowe Jr using Paint Shop Pro X7.

Joe Wisinski edited and proofread the manuscript.

http://www.linkedin.com/in/joewisinski

The manuscript was reviewed by Bob W. Locke, Ken Cureton, Jimmy James and Steve Levinson. Thanks to Mardhavi Sakuntala for helping with promotion.

You can find up-to-date information at the Home Computer Security blog at http://www.leave-me-alone.com

This book is published by The Writing King. Paperback version is published via Createspace. Kindle version is published via Kindle Direct Publishing.

The Writing King: http://www.thewritingking.com
The Ghost Publishing: http://theghostpublishing.com

Other web sites operated by Richard Lowe Jr:
Personal website: http://www.richardlowe.com
Photography: http://www.richardlowejr.com
LinkedIn Profile: https://www.linkedin.com/in/richardlowejr

If you have any comments about this book, feel free to email Richard G. Lowe Jr at rich@thewritingking.com

www.ingramcontent.com/pod-product-compliance
Lightning Source LLC
Chambersburg PA
CBHW081810200326
41597CB00023B/4215